Sustainable Transportation Options for the 21st Century and Beyond

C.E. (Sandy) Thomas

Sustainable Transportation Options for the 21st Century and Beyond

A Comprehensive Comparison of Alternatives to the Internal Combustion Engine

Springer

C.E. (Sandy) Thomas
H2Gen Innovations, Inc. of Alexandria
Alexandria, VA
USA

ISBN 978-3-319-16831-9 ISBN 978-3-319-16832-6 (eBook)
DOI 10.1007/978-3-319-16832-6

Library of Congress Control Number: 2015943365

Springer Cham Heidelberg New York Dordrecht London

Printed on acid-free paper

Springer International Publishing AG Switzerland is part of Springer Science+Business Media (www.springer.com)

Foreword

Fundamental change of any major system in democracies takes a long time. It is the nature of public participation in a representative or self-governing system that options for change are examined carefully and thoroughly so that "the people" can ultimately decide what is best for them. We suspect and therefore reject top-down decisions made by a few for the many, especially in the USA. We tend to think that there is more in it for them than for us. We question their motives and their knowledge; we doubt their solutions; we tend to think the worst of their rationale for change as we protect and prolong our status quo.

It means that our democratic ways are messy, confusing, uncertain, and for those who are best positioned and knowledgeable, who in fact actually do know more than most, quite frustrating. Such people are the agents of change and most often its advocates and true believers, adding burden to the weary task of waiting for others to catch their own knowledge and awareness up to levels sufficient to support such change. So it has been, is now, and likely will continue to be when it comes to determining and implementing our personal transportation future in the twenty-first century.

We start with a dedication to what we know works: the current system. When it comes to current US personal transportation, how could we be wrong with well in excess of 200 million internal combustion vehicles on the road today? There are hundreds of millions more vehicles, just like US vehicles, across the rest of the world. New car sales number in the tens of millions annually. They carry us in comfort and safety to wherever we want to go. They are reliable and, as far as we know, efficient, delivering for most drivers more than 300 miles of transportation between fuel refills. They function relatively cleanly, or at least cleaner than they were decades ago. Most of the countries where they operate enjoy clear blue skies unless there is a weather front, or clear blue between broken clouds. They are also affordable, given the choices and financial packages available to purchase or lease them. And they are long lived, lasting for well more than a decade in most cases, replaced more often by preferential choice than need.

We have also spent many decades establishing the infrastructure to enable these vehicles to serve us. Apart from those who are themselves handy do-it-yourselfers when it comes to maintenance and repair, refueling, and cleaning, we have an entire fueling, repair, maintenance, and cleaning system to take care of our cars for us. Their employees are expert at what they do, understanding the technologies and mechanics they support and often making a life's work of it. Moreover many national economies benefit from the commercial aspects of designing, manufacturing, distributing, selling, and supplying components and subassemblies for cars as they are, as well as the assembled product. Why would we seek to actively disrupt or even disturb this manufacturing and supply system and infrastructure or threaten it with unwanted or unrequested changes that they themselves do not originate for the sake of a new power system under the hood?

Inertia is a powerful force, most often underestimated by change agents, especially when what you have got is perceived as a good thing that is working well. "Don't fix what ain't broke" is advice from the ages. Thought leaders, individuals who head corporations, institutions, societies, and nations, know full well from experience that they cannot lean too far in new directions before resistance sets in. Unless they are prepared and geared up for the struggle they will face to make the change they choose to pursue, they may more likely fail than succeed. History records more failed change efforts than successes for this very reason. What are we doing then challenging the existing personal transportation system that works so well and why is our author proposing fundamental change to the power source that ultimately propels the mobility that we so enjoy and depend upon in our contemporary lives? Why change and why change now?

The keyword for the answer has everything to do with sustainability: the future life of the planet and its billions of inhabitants. The current power source is a remarkable machine that has served us well. It is an historical marvel to have done so well for so long. But it hails from a time when we knew what we knew then but comes nowhere close to what we know now about so many things associated with this power source and its fuel and the alternatives. Yes, it is a remarkable machine, more complex than most even understand. It is also within the laws of thermodynamics quite inefficient from the point of combustion in the engine chamber to the spot on the pavement where power becomes force thanks to the ultimate turn of the wheels that push the vehicle. It is also fueled by a finite fuel source that has been proven to be dangerous in ways that we did not know, could not see, and had no idea of its cumulative impact on the risks to safety, health, and life itself. More recently, the fuel has also become both a financial hardship and an ever greater international and geopolitical risk to relations among nations. Simply put the device, that is the engine, is replaceable and its fuel produces more risks than technically and readily available alternatives. This simply drives the conclusion among more and more people from every society around the world that it is time for a change. It is time for a complete replacement of the personal transportation power source and its fuel, in fact.

We have the knowledge, the technology, the manufacturability, the wherewithal, and the capacity to implement a totally new solution. It takes advantage of current knowledge, meets and exceeds our power needs, is increasingly affordable, especially when high volumes impact unit costs, utilizes an infinite fuel source, and produces no pollution. Given the inherent goodness of the alternative, why are we not madly rushing forward to take advantage of this new device in pell-mell fashion, with all haste pushing it? That is certainly an option with what we know now. The reason is we lack the will. Our democratic dialogue is not yet far enough along. We lack the information we need that this book provides.

Let us remember. In a democracy, there are many voices, opinions, and indeed there are also other options. They will all be heard from. We are entering, let us hope, the final stages of the discussion and debate before we get on with implementing the "final solution" that Sandy Thomas describes so completely and well.

To make the ultimate transition however, we must go through the last stages of the dialogue. This will include a more comprehensive understanding and appreciation of the hydrogen fuel cell and its capabilities, the requirements of the infrastructure to enable and support it, and just as importantly further conversation and consensus of the rest of the systems changes needed to bring it all together. In this latter regard, our multi-faceted transportation system and its final conversion to fuel cell power systems will have to answer the questions raised within the legal, social, economic, and political systems supporting it.

These are indeed critical questions that must satisfy myriad stakeholders and most importantly the manufacturers of the vehicles, the providers of the infrastructure, and ultimately the consumers who will purchase, use, and enjoy their new mobility. I have no doubt the questions are answerable, and frankly, most of them are thanks to the explanations and descriptions in this book. However, we will be thorough in creating the pathways to our new mobility and so whatever time it takes to satisfy the needs of the populace, writ large, that's how long it will take to reach the ultimate tipping point to HFC mobility.

The legal system enablers are of course critical in nations and a world of laws. Risks must be addressed and satisfactorily covered. The production, distribution, and storage of hydrogen need safeguards and thus regulations to insure safety. Compliance and enforcement must be assured as well. Our social understanding of hydrogen and fuel cells has to get beyond the tragedy of the *Hindenburg* and other anecdotes of disaster and tragedy. We can do this with proper education, communication, and practical experience. The myths versus facts in the economics of fuel cells and hydrogen must be understood for what they are and which is which. We could never have gotten to where we are today without those who fully believe in and have committed themselves to the commercial success of this product, its fuel and their combination in our vehicles. We know the environmental opportunity that the conversion creates and many who support this technology are "chomping at the bit" to realize such a transformation for permanently cleaner air, water, and land.

What is inevitably holding back the possible are the politics of the systems changes. This is where the forces of the status quo, our current legacy transportation system and subsystems, those who are invested in what is today and their tremendous sway over those who advocate for change, slow and deter the political process to suit themselves. It is their right of course as it is the right of all of those who promote a different form of change, ranging from alternative fuels for internal combustion engines to the battery electric vehicle as an alternative to the fuel cell electric vehicle. As said previously, democracy is messy and often chaotic. I have no doubt however that we will get to where we need to go, thanks in large measure to the clear science and unambiguous description and explanations that our author provides us. Politics is a dynamic and moving feast. Its propensity for delay and debate is only exceeded by its ultimate pragmatism in the face of clear and present reality. The internal combustion engine, fueled by oil products or ultimately any alternative fuel, lacks the efficiency and effectiveness of the hydrogen fuel cell vehicle. The battery electric vehicle cannot ever overcome its inherent disadvantages to the hydrogen fuel cell. So we will get there in the end. Politics will come round. Proponents and adherents will learn more and say more, influencing and persuading ever more people anxious to both clean up and lower their costs of transportation. Mobility is an ever more important part of civil society and life in the twenty-first century. We will ultimately want the best for ourselves and demand that our political system deliver exactly that.

When we finally get there, we will all owe an enormous debt of gratitude to our author, Sandy Thomas, for his foresight, persistence, forbearance, research, development, study, common sense, and commitment. He has been a pioneer, advocate, and implementer of the science, technology, and tools to deliver sustainable transportation to the democratic masses. We are fortunately coming ever closer to the tipping point of the change that he has sought and promoted for many years. We will all be the better for it when and as it comes. Clean, reliable, affordable, powerful, and sustainable transportation for the twenty-first century is around the corner. Early commercial vehicles are already on the roads of Japan, Europe, and the USA . They are the misty droplets before the steady rain. This book explains what is coming, why, and what benefits will become us.

John Hofmeister
Chair—US Department of Energy Hydrogen Technology and Fuel Cell Advisory Committee (2012–2015)
Founder and CEO—Citizens for Affordable Energy, Inc.,
Retired President—Shell Oil Company

Acknowledgments

My interest in hydrogen-powered fuel cell electric vehicles was first stimulated by the visionary work of Professor Joan Ogden.[1] I was first introduced to hydrogen and fuel cells by Dr. Ogden's ground-breaking 1989 123-page booklet for the World Resources Institute with Robert Williams from Princeton entitled "Solar Hydrogen: Moving Beyond Fossil Fuels." At the time I was a legislative assistant to Senator Tom Harkin of Iowa. Senator Harkin accelerated the development of hydrogen as an energy carrier when he succeeded in doubling the early hydrogen research budget at the Department of Energy from $1 million to $2 million. The DOE's hydrogen budget subsequently rose to over $200 million per year, but it was Senator Harkin's influence and that of Representative Bob Walker of Pennsylvania that kept the Federal hydrogen energy research budget growing through congressional appropriations in the early years of hydrogen development.

I would also like to thank: Dr. Bob Shaw, and the other investors in H_2Gen Innovations, Inc. of Alexandria, Virginia. Dr. Shaw was the lead investor in H_2Gen, the company we set up to develop and commercialize steam methane reformers to convert natural gas to hydrogen at the fueling station.

Ira Kuhn, Jr., was then the President of Directed Technologies, Inc. (DTI) of Arlington, Virginia. Ira recognized the potential of hydrogen as an energy carrier early on and convinced the Ford Motor Company in 1993 to undertake a joint development with DTI for the US DOE to explore hydrogen-powered FCEVs and the means to safely deliver affordable hydrogen to those vehicles.

Dr. Sigmund Gronich led the hydrogen and fuel cell development programs at the US Department of Energy for many years; Dr. Gronich read and provided valuable advice and suggested improvements to early drafts of this book.

Brian D. James has been conducting in-depth mass production cost estimates of fuel cells and hydrogen energy equipment for the Ford Motor Company and the

[1] In 1989 Dr. Ogden was a Professor at Princeton University; she has since moved to the Institute for Transportation Studies at University of California at Davis where she continues her research on hydrogen and fuel cells, training a new batch of young students on the merits of hydrogen as an energy carrier in the process.

Department of Energy when he worked at DTI, continuing today with Strategic Analysis, Inc. Brian made many helpful comments on early drafts of Chap. 13 and Appendix E.

Dr. Frank Lomax, a brilliant engineer, was the backbone of the scientific and engineering efforts at both DTI and H_2Gen to develop affordable hydrogen energy systems.

Ron Sims and Brad Bates then at the Research Department of the Ford Motor company in Dearborn, Michigan, helped set up and run the DOE/DTI/Ford hydrogen and FCEV Development program.

Bob Rose, the leader of the Fuel Cell Council, an early industry association dedicated to the development of fuel cell systems, and Jeff Serfass, the former President of the National Hydrogen Association (NHA),[2] set up to promote the commercial uses of hydrogen as an energy carrier.

Frank Novachek with Excel Energy, a large Midwestern utility; Frank led the group development of several explorations of the merits of hydrogen and fuel cells for the NHA.

[2]The Fuel Cell Council and the NHA were subsequently merged in 2010 to form the Fuel Cell and Hydrogen Energy Association a trade association led by Morry Markowitz.

Contents

Abbreviations

ADG	Anaerobic Digester Gas
AEO	Annual Energy Outlook (DOE publication)
AER	All Electric Range
AFV	Alternative Fuel Vehicle
ANL	Argonne National Laboratory
bbl	Barrel
BEV	Battery Electric Vehicle
CARB	California Air Resources Board
CCS	Combined Cycle
CCS	Carbon Capture and Sequestration (or Storage)
CF	Capacity Factor
CO_2	Carbon Dioxide
DFMA	Design for Manufacturing and Assembly
DOD	Depth of Discharge
DOE	(US) Department of Energy
DTI	Directed Technologies, Inc.
E-85	Mix of 85 % ethanol in gasoline
EIA	Energy Information Administration
EPA	(US) Environmental Protection Agency
EtOH	Ethanol
EU	European Union
EV	Electric Vehicle
FCEV	Fuel Cell Electric Vehicle
GH2	Gaseous Hydrogen
GHG	Greenhouse Gas
GREET	Greenhouse gases, Regulated Emissions, and Energy use in Transportation (Argonne National Laboratory)
H_2	Hydrogen
H2A	Department of Energy Hydrogen Cost Program
HEV	Hybrid Electric Vehicle
ICE	Internal Combustion Engine

ICV	Internal Combustion (engine) Vehicle
IGCC	Integrated Gasification Combined Cycle (Coal power plant)
IPCC	Intergovernmental Panel on Climate Change
LDV	Light Duty Vehicle
LH2	Liquid Hydrogen
Li-Ion	Lithium Ion (battery)
MCFC	Molten Carbonate Fuel Cell
MIT	Massachusetts Institute of Technology
NETL	National Energy Technology Laboratory
NG	Natural Gas
NGV	Natural Gas Vehicle
NiMH	Nickel Metal Hydrid (battery)
NREL	National Renewable Energy Laboratory
NT SUV	Non-Truck SUV
OEM	Original Equipment Manufacturer (usually Car companies)
OPEC	Organization of Petroleum Exporting Countries
Pba	Lead Acid (battery)
PEM	Proton Exchange Membrane (type of fuel cell)
PHEV	Plug-in Hybrid Electric Vehicle
PSA	Pressure Swing Adsorption
RMI	Rocky Mountain Institute
SOFC	Solid Oxide Fuel Cell
SMR	Steam Methane Reformer
SUV	Sport Utility Vehicle
USABC	US Advanced Battery Consortium
VMT	Vehicle Miles Traveled
WWTP	Wastewater Treatment Plant
ZEV	Zero-Emission Vehicle

Chapter 1
Introduction

Transportation dominates environmental and energy security threats to the world. Light-duty vehicles (LDVs—cars and trucks) accounted for 45.6 % of all the oil consumed[1] by the USA in 2009 and 37.4 % of all greenhouse gas (GHG) emissions.[2] Assuming that economic growth resumes after the 2008–2009 global recession, US vehicle sales and annual miles traveled per vehicle will most likely resume their historical upward trajectories (although probably not as steeply) which will lead to a steady increase in urban air pollution, GHG emissions, and oil consumption, unless we change the way we power our cars, trucks, buses, planes, trains, and ships. Our petroleum consumption and pollution from vehicles in the USA is exacerbated by long travel distances, minimal mass transit, and very low fuel taxes compared to the rest of the world.

Change in our modes of transportation is crucial for the USA and for the world. If the rest of the world adopts our dirty habits of burning billions of gallons of oil to propel vehicles for their rapidly expanding and increasingly mobile populations, then GHG emissions and local air pollution will become intolerable and oil consumption unsupportable given the world's dwindling supplies of crude oil.

The venerable internal combustion engine (ICE) must be replaced. One or two generations from now, children will be incredulous when their grandparents attempt to describe how ICEs operated.

Grandchild:

> Do you mean to say that they actually set off multiple *explosions* in six or eight cans that each contained a moving projectile, and they attached each of those projectiles to a connecting rod to convert linear motion of the projectiles into rotary motion to drive the wheels? That sounds noisy, dangerous and bizarre!

[1]Total US transportation (including all heavy-duty trucks, planes and ships in addition to LDVs) consumed 71 % of all oil, but LDVs accounted for 64 % of the total transportation oil consumption for a net consumption of $0.64 \times 0.71 = 45.6$ %.

[2]All transportation vehicles generated 57.3 % of all US GHGs in 2009, of which LDVs accounted for 65.2 % or $0.573 \times 0.652 = 37.4$ %.

C.E. (Sandy) Thomas, *Sustainable Transportation Options for the 21st Century and Beyond*, DOI 10.1007/978-3-319-16832-6_1

Indeed, it is a tribute to automotive engineers over the twentieth century that ICEs operated quietly, relatively safely, and lasted through so many thousands of detonations with minimal repair cost.

But, society can no longer afford to use ICEs to propel our vehicles. Even the best ICE running on diesel fuel has a peak efficiency of 42 %, with actual on-the-road efficiencies of only 10–12 % averaged over typical drive cycles. Most of the energy in the gasoline or diesel fuel is converted to heat, which may help keep the passengers warm in the winter, but wastes most of the energy from burning the gasoline or diesel fuel. For comparison, the likely replacement for the ICE, an electric motor, has efficiencies of 80–90 % or more independent of the drive cycle.

The electric motor also has the advantage of delivering high torque at zero velocity. The ICE, on the other hand, must ramp up to relatively high speed before it generates adequate torque to move a vehicle, which dictates the need for a complex and expensive mechanical transmission to couple a high-speed engine to a stopped or slowly moving vehicle. No transmission is required for an electric motor that generates high torque or force even at standstill (except possibly to switch from forward to reverse motion, a much less complex transmission).

Finally, electric motors are much more durable than ICEs. With only one moving part, electric motors used in delivery trucks in the UK have routinely exceeded 100,000 miles operation without overhaul.

Thus, electric motors are quieter with better acceleration and better durability than the ICE.

While electric motors must replace ICEs to solve our long-term environmental and energy security threats, there are several different options to provide the electricity to run those motors.

To most people, an "electric vehicle" or "EV" implies a vehicle powered by batteries, which we designate here as a battery electric vehicle or battery EV or BEV. But, electricity can also be generated onboard a vehicle by coupling an ICE to an electrical generator to recharge the battery, which we now call a hybrid electric vehicle or HEV. Alternately, electricity can be supplied to the motor by a fuel cell that is typically fueled with hydrogen, a fuel cell electric vehicle (FCEV) or fuel cell EV.

Finally, the batteries on a hybrid electric vehicle[3] can be charged by grid electricity…a plug-in hybrid electric vehicle or PHEV. General Motors has been selling a PHEV since December 2010, the Chevy Volt, which GM calls an extended-range electric vehicle or E-REV (we use the PHEV designation in this book for all plug-in hybrids).

Electrification of the motor vehicle fleet has already begun with the introduction of hybrid electric vehicles (HEVs) such as the Honda Insight and Toyota Prius.

[3]Or on a fuel cell EV; hydrogen-powered fuel cell EVs also have batteries to increase peak power and to allow the capture of energy while braking, called "regenerative braking," and these batteries can be charged from the grid at night; the very first plug-in electric vehicle by a major auto company was a prototype plug-in fuel cell EV made by the Ford Motor Company on their Edge platform in the early 2000s.

Hybrid EVs typically have two modes of power: an electric motor and a small ICE that can be coupled to the wheels through a planetary gear and can also be used to drive a generator to recharge car batteries. Through the end of 2012, total HEV sales in the USA from all the automobile companies reached 2.54 million [1]. A total of 382,704 HEVs were sold in 2012, which was equal to 4.5 % of all cars sold in the USA that year. Total global sales of hybrid EVs reach approximately 4.8 million.[4]

This book provides detailed comparisons of these electric vehicles with respect to:

- Initial vehicle cost;
- Fuel infrastructure cost;
- Market Penetration potential;
- GHG emissions;
- Local air pollution; and
- Oil consumption.

We also include an analysis of three other fuels: biofuels, natural gas, and diesel fuel. All three fuels can be used to power an ICE, so these are not fuels for electric vehicles.[5] We assume in all cases that these alternative fuels are used in HEVs or PHEVs, where they provide the best benefits in terms of reduced environmental footprint and lower oil consumption.

After studying all the options, we conclude that the hydrogen-powered fuel cell EV is superior in all six attributes listed above compared to the alternatives. Fuel cell EVs generate lower GHGs, consume less oil, are projected to cost less in mass production, and require lower cost fuel infrastructure than battery EVs, plug-in hybrid EVs, or other alternatives.

Hydrogen and FCEV programs are being developed aggressively around the world; Germany, Japan, S. Korea, Scandinavia, and the UK and are actively supporting robust hydrogen and fuel cell EV programs on the premise that fuel cell EVs offer the best prospect of meeting long-term environmental and energy security goals (see Chap. 12). The USA is now losing its former premier position in the hydrogen and fuel cell EV arena. The USA risks falling behind once again, just as it fell behind in the development of gasoline hybrid EVs. In the case of HEVs, the USA could later import hybrids from Japan to improve light-duty

---

[4]This estimate is based on total cumulative hybrid EV sales of 2.18 M in the USA; 1.5 M in Japan, and 0.45 M in the EU through September 2012, which is 1.89 times the USA cumulative hybrid EV sales by that time.

[5]Although in principle biofuels or natural gas could be converted to hydrogen onboard a vehicle with a small chemical plant to power fuel cell EV, onboard reforming was attempted in the 1990s, particularly for methanol which can be reformed into hydrogen at lower temperature than other fuels, but most auto companies have now concluded that onboard reforming is too expensive and hydrogen will most likely be stored as a compressed gas on fuel cell EVs, and all FCEVs being sold, leased, or tested on the road today use compressed hydrogen as the primary fuel.

vehicle fuel economy and still reap the benefits of lower pollution and reduced oil consumption here in the USA. This may not be possible in the short term with fuel cell EVs, since they require a distributed hydrogen fueling infrastructure, which other nations are now installing. The USA will not be able to import fuel cell EVs without the necessary hydrogen infrastructure in place. Ironically, as shown in Chap. 6, the money spent by the administration on electrical outlets to charge battery EVs would allow the installation of hydrogen fueling stations that could support over five times more fuel cell EVs per tax dollar spent than the battery EVs and plug-in hybrid EVs supported by electrical charging outlets.

Fortunately various US states have active programs to develop and deploy hydrogen infrastructure and FCEVs. California, Hawaii, New York, Connecticut, South Carolina, and Ohio among others have active hydrogen and fuel cell programs. If they are able to maintain their hydrogen and fuel cell EV programs despite their budget deficits, then they might provide the necessary infrastructure to jump-start a robust introduction of fuel cell EVs in the USA in the coming decades when the limitations of battery EVs and plug-in hybrid EVs relative to fuel cell EVs become more apparent and the need to curb GHGs more urgent. California has an aggressive program to install a robust hydrogen infrastructure so that the automobile companies can begin selling FCEVs in California in the 2015–2020 time period.

Fortunately, most major car companies did not wait for the US Federal Government to pursue the development of fuel cell EVs. Many car companies have acknowledged that the fuel cell EV will most likely be the ultimate alternative vehicle in terms of curbing environmental abuse and curbing our addiction to foreign oil. While the Bush administration invested $1.2 billion of taxpayer dollars in hydrogen and fuel cell EV development, it is estimated that the car companies collectively invested $7 to $8 billion of private capital to develop this technology over the same time period. Several car companies have announced their intention to mass produce hydrogen-powered fuel cell EVs by 2015–2018. The question is whether these zero emission vehicles will be sold only in Germany, Japan, S. Korea, the UK and Scandinavia, or will the USA provide a robust environment for FCEVs?

One exciting alternative to burning fossil fuels to propel out automobiles is to utilize waste products to generate hydrogen as the fuel to power FCEVs. By running FCEVs on hydrogen generated from solid waste in our nation's many municipal wastewater treatment plants (WWTPs) or from landfill gas as described in Chap. 13,[6] we eliminate the need for fossil fuel imports, eliminate local air pollution, and generate zero or negative[7] (GHG) emissions; however, the hydrogen that

[6]Hydrogen for fueling FCEVs is already being generated from waste at the Orange County Sanitation District WWTP in Fountain Valley, California.

[7]GHG emissions could be *negative* if the hydrogen was made from the anaerobic digester gas (ADG) that is emitted from existing municipal waste water treatment plants. Methane, a gas that is 25 times more potent than carbon dioxide as a GHG, is sometimes vented or flared from WWTPs. Converting this wasted gas to hydrogen for use in (FCEVs) would decrease GHG emissions compared to current practice.

could be produced from (WWTPs) and landfill gases would support less than 2.6 % of all US (LDVs).

In the long term, most hydrogen will have to be produced from other sources such as:

- Nuclear power and renewable energy used to electrolyze water to produce hydrogen.
  - Advantages: zero carbon pathway; hydrogen can be stored in the case of intermittent renewables to provide a steady source of renewable electricity through stationary fuel cells
  - Disadvantages: uses freshwater that is in short supply in many parts of the world, but at least one company is developing an electrolysis option that uses seawater.
- Ethanol
  - Advantages: Ethanol distribution infrastructure already exists in many parts of the USA.
  - Disadvantages: Most ethanol is currently made from corn grain in the USA, but cellulosic ethanol can also be made from non-edible plants including corn stover[8] and switchgrass, farmed trees, among others.
- Coal through a high efficiency integrated gasification combined cycle (IGCC) plant plus carbon capture and sequestration (CCS)
  - Advantages: The reserves of coal are huge in the USA and would alleviate the world's dependence on petroleum imported from the volatile Middle East or OPEC nations.
  - Developing coal gasification with carbon capture and storage (CCS) would provide a clean pathway to produce electricity.
  - Coal gasification plants could be used to produce hydrogen and also to provide electricity via stationary fuel cell systems.
  - Carbon capture is easier with precombustion separation of the CO_2 in an IGCC plant than with post-combustion attempts to recover the diluted CO_2 from the smokestack of a conventional coal boiler plant.
- Non-edible biomass

Reference

1. Hybrid electric vehicle sales, available at http://www.afdc.energy.gov/data/categories/afvs-and-hevs

[8]Stover includes the non-edible roots and stalks of the corn plant.

Chapter 2
Societal Requirements/Goals for a Sustainable Transportation System

Abstract This chapter defines our goals for a future transportation system in terms of the allowable greenhouse gas (GHG) emissions, fossil fuel consumption, and local air pollution to create a sustainable transportation system.

Every engineering project (in this case developing alternative transportation systems) needs to have measurable goals so that one can judge the success of a given design or designs.

Engineers typically include a set of *boundaries* to help to define the system that they are trying to evaluate. If they are designing a new car, the boundaries might encompass the entire car, with key inputs to the system being the quantity of gasoline or diesel fuel needed to travel 100 miles, and key outputs would include the gases exiting the tailpipe plus any fuel evaporation and particulate matter (PM) generated by applying the brakes and particles released due to tire wear on the road. But to analyze an entire transportation system, the boundaries must be expanded to include all inputs to the entire chain of manufacturing the vehicle and producing the fuel to power that vehicle. And the analysis must include all emissions from all vehicle manufacturing and fuel production operations as well as the actual emissions from the vehicle. One key example of the need for large boundaries is the greenhouse gas (GHG) emissions caused by driving a battery electric vehicle. The BEV itself emits no GHGs. But three quarters of all power plants in the USA use fossil fuels to generate electricity to charge car batteries. These power plants emit copious quantities of GHGs, primarily carbon dioxide (CO_2), the main GHG, and must be counted in the evaluation of the global impact of driving battery EVs; as more BEVs travel longer distances, GHG emissions from the electrical power plants will increase.

In other words, to properly evaluate the performance of a transportation system, we must consider civilization as a closed factory system that is supplied with three non-renewable fossil fuel tanks to supply most of its energy (petroleum, coal, and natural gas) and one renewable energy source (the sun) plus nuclear energy. This vast factory system is enclosed by a tent (called the atmosphere) that has a finite

C.E. (Sandy) Thomas, *Sustainable Transportation Options for the 21st Century and Beyond*, DOI 10.1007/978-3-319-16832-6_2

capacity to absorb waste products (primarily air pollution and GHG emissions). To continue sustainable operation, this factory must conserve the energy in its dwindling non-renewable fossil fuel tanks, and it must limit its waste streams to avoid poisoning the planet to such a degree that the health of future generations is put at grave risk.

This global factory system has four key sections: industrial, commercial, residential, and transportation. All these societal sections must conserve energy, water, and other material resources and minimize pollution to maintain a sustainable factory for future generations; however, this book focusses on the transportation sector, which consumes much of the world's energy and generates much of its pollution. In order to preserve anything close to our current standard of living in this global tent, we must set a goal of limiting at least three parameters of the transportation sector: GHG emissions, oil consumption, and local air pollution.[1]

2.1 Greenhouse Gas Emission Requirements

Transportation accounted for 34 % of all GHG emissions in the USA in 2012, and light-duty vehicles (LDVs) (cars and trucks) generated 61.5 % of those transportation GHGs,[2] so cars and light-duty trucks accounted for 21 % of all US GHGs [1]. Since GHGs (primarily CO_2) stay in the atmosphere for a century or more,[3] we must stop spewing these gases soon if we are to avoid the accumulation of excessive gases that could dramatically impact the future of our children and grandchildren. The pollution that we pour into the atmosphere today will still be plaguing our grandchildren and great-grandchildren 100 years from now.

The goal of the environmental community to avoid the most serious effects of climate change[4] is to decrease GHGs by a least 80 % below 1990 levels, ideally by 2050. In 2012, US transportation GHGs had increased by 18 % since 1990, so if transportation were required to cut GHGs 80 % below its 1990 level, that would require an 83 % reduction below 2012 levels. US industrial GHG emissions have already decreased by 10.8 % since 1990, so it may be difficult to cut industrial emissions further. It may also be more difficult to cut residential GHG emissions since the housing stock turns over slowly, so it may be necessary to require more

[1]Clean water should probably be added to this list, since water is becoming a scarce resource in many parts of the world. All else being equal, we should choose transportation systems (including fuel production) that minimize the consumption of water.

[2]Light-duty trucks accounted for 18.4 % of all transportation GHGs in 2012 and light-duty cars 43.1 %.

[3]CO_2 has an atmospheric lifetime of approximately 100 years, while the other major GHGs such as methane (CH_4) and nitrous oxide (N_2O) have lifetimes of 12 and 298 years, respectively.

[4]The goal of the IPCC (Intergovernmental Panel on Climate Change) is to keep the global average temperature rise to no more than 2 °C (or 3.6 °F) above preindustrial levels.

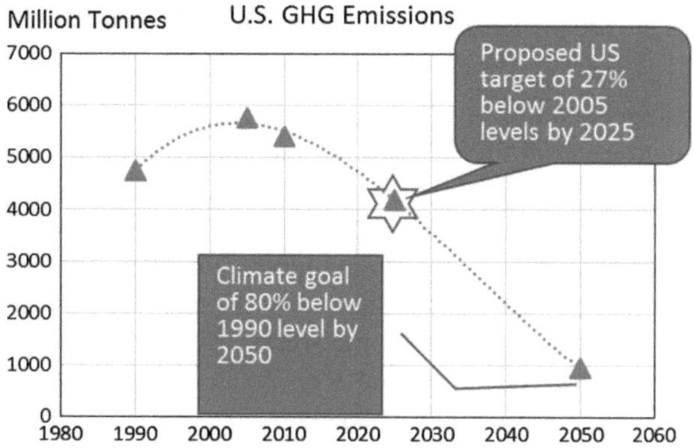

Fig. 2.1 US GHG emissions and the climate change goal of an 80 % reduction of GHGs by 2050 compared to the proposed goal of 26 % to 28 % below 2005 levels by 2025

than an 80–83 % reduction for transportation to accommodate other sectors missing the 80 % reduction goal.

For example, in a detailed analysis of GHG emissions in the European Union, McKinsey & Company concluded that road transportation would have to be *reduced by* 95 % below 1990 levels, after taking into account the likely reductions possible in other sectors of the EU economy [2]. For example, they concluded that air and sea transports could only be reduced by 50 % and industrial GHGs by 40 % and agriculture by 20 %, leaving a larger burden of 95 % reduction for road transport.

Assuming that medium- and heavy-duty trucks are included in "road transport" that accounts for 20 % of all US GHGs, then one could argue that LDVs would have to be reduced by more than 95 % below 1990 levels to reach a total reduction of 95 % for all road transport.[5]

On a trip to China on November 12, 2014, President Obama and Chinese President Xi agreed to set a goal of reducing GHGs by 26–28 % below 2005 emission levels by 2025 [3]. This is a worthy step in the right direction, but is only 11.5 % below the 1990 US GHG emission level, which might seem far short of the 80 % reduction goal. However, this 11.5 % reduction by 2025 does fall on the glide slope[6] toward the climate change community goal of an 80 % reduction below 1990 levels by 2050 as shown in Fig. 2.1, even though the US 2005 GHG level was 21 % above 1990 GHG emission levels.

We conclude that the GHG reduction goal for LDVs should be at least an 80 % reduction and preferably a 95 % reduction below 1990 levels, preferably by 2050, but at least by the second half of the twenty-first century.

[5]This assumes that it will be more difficult to reduce GHGs from medium- and heavy-duty trucks than from light-duty vehicles, which would certainly be true for battery EVs, but not necessarily for fuel cell EVs. To the degree that hydrogen and fuel cells could also power heavy trucks, and air and sea transports, the GHG reduction targets for light-duty vehicles might be reduced.

[6]A cubic polynomial fit to the data.

2.2 Dependence on Finite Fossil Fuels

Some US citizens may becoming less anxious about petroleum dependence on nations from OPEC with the precipitous drop in oil prices in 2014 and 2015 coupled with new oil production in the USA through fracking; but OPEC nations still have more than 72 % of proved global petroleum reserves (see Fig. 2.6) and could therefore pose a significant global threat in some future crisis unless developed nations reduce our consumption of petroleum. Most citizens think of motor vehicles when they consider the world's voracious and growing appetite for petroleum. Transportation accounted for 72 % of all petroleum consumption in the USA in 2013 as shown in Fig. 2.2 [4].

Of this transportation amount, LDVs accounted for 58.6 % [5] (see Fig. 2.3), so that LDVs consumed 42.4 % of all US petroleum in 2013.

But petroleum use is much more pervasive than just in transportation. As UCLA Professor Jared Diamond hypothesized [6] when he explored the reasons why modern civilization grew and flourished in Asia and Europe (and subsequently in the USA), but not in places that he analyzed like Papua New Guinea,

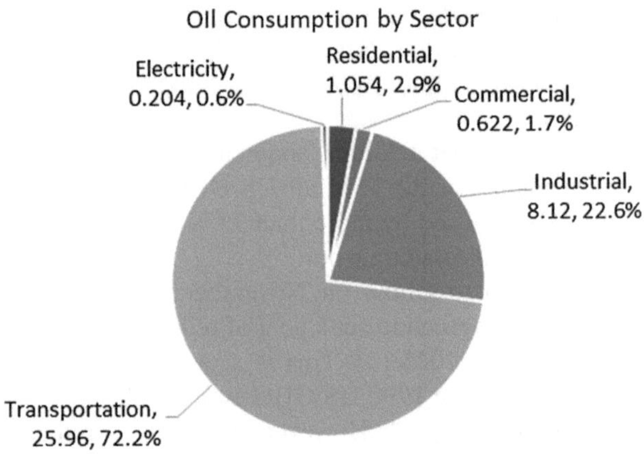

Fig. 2.2 US oil consumption (quads) by sector

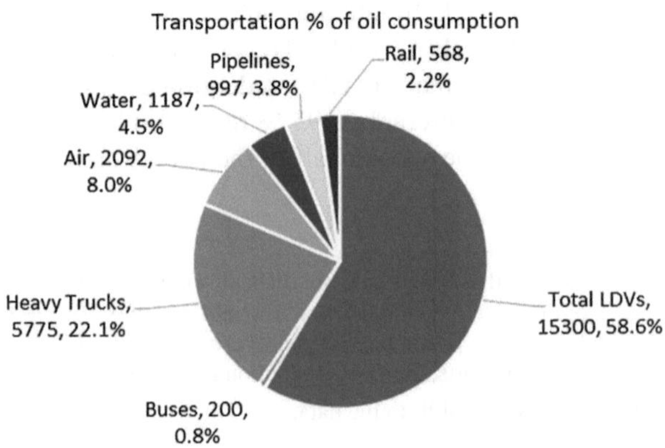

Fig. 2.3 Transportation sector % oil consumption by mode of travel (trillion BTUs) for 2013

where natives in the highlands still live in a prehistory "hunter-gatherer culture," modern civilizations developed where two key resources were available or could be successfully imported: domesticated, high-yield crops such as wheat, barley, and rice and domesticated animals such as pigs, sheep, cattle, and horses (and particularly cattle and horses that provided mechanical force to augment human farm labor). These two ingredients allowed ancient hunter-gatherer peoples in the Fertile Crescent (Mesopotamia and now modern Iran and Iraq) 12,000 years ago to settle down and form the first cities since farmers could grow enough food to support large groups of people. Prof Diamond concluded that the people he met in Papua New Guinea were just as smart and probably more resourceful and hard-working than American or European citizens, but they lack the climate that could support high-yield crops or horses or cattle to help with farming chores.

When the first Europeans brought cereal grains and horses to the USA, and American farmers moved to the fertile Midwest of the USA, the essential ingredients for a thriving civilization were planted. The advent of the steam engine and later the internal combustion engine allowed farmers to improve their yields many times over compared to relying strictly on horses to power farm equipment. While the end result is striking in terms of food production capacity, food production is now highly dependent on petroleum for everything from fertilizers, pesticides, herbicides, and fossil fuels for farm equipment, and gasoline and diesel fuel to transport farm produce to the food processing plant, and still more petroleum products to bring the finished food products to the supermarket. In 1994, one study estimated that growing food consumed *400 gal of petroleum to produce the food for one American* [7]. This amount of gasoline would propel a vehicle with 30 miles/gal fuel economy 12,000 miles/per year.

So what should be our goal for reduction of oil consumption? US politicians of all stripes have called for "energy independence" for many decades, setting the goal of eliminating the importation of *any* foreign energy supplies.[7] While this makes great campaign rhetoric in terms of offering the promise of cutting the military costs for defending our oil lifeline to the volatile Middle East and offering the chimera of reduced gasoline prices at the pump, in reality the goal of energy independence is probably not attainable in the foreseeable future and would probably not yield the promised advantages. Energy independence will most likely not be achievable for many decades, if ever, since the USA would have to greatly expand the recent surge in shale oil production and probably develop huge quantities of alternative energy sources such as wind, solar, biomass, coal gasification, or possibly nuclear fusion. Ironically, any of these alternative energy sources would most likely require a huge surge in oil consumption to manufacture and install these new systems and the investments of tens of billions of dollars by some combination of industry and government to build new infrastructure. In terms of reduced gasoline prices (or alternative fuels such as electricity, hydrogen or biofuels), most

[7]Denmark set a goal of eliminating all fossil fuel energy use in transportation by 2050, relying primarily on hydrogen-powered vehicles [8].

of these new energy sources will be more expensive than gasoline per unit of energy, at least in the short term.[8]

But even if the USA were somehow able to achieve energy independence, we live in a global economy, as dramatically illustrated by the impact of the European debt crisis on US stock markets in the deep recession in the 2010–2011 time period. Unless all nations could duplicate our energy independence feat, the world would still depend on oil coming from the Persian Gulf, and OPEC would still be in a position to disrupt the global economy. So we would still be gravely concerned about any unrest in the Middle East that could disrupt the flow of oil to our friends and allies.

Society has not established a specific target for reducing petroleum consumption analogous to the 80 % reduction goal for GHGs, but one goal might be to reduce oil consumption such that in some future emergency, the USA could supply all of its petroleum needs from non-OPEC nations and, preferably, from friendly countries on the American continent.

Fortunately, these goals seem attainable. In 2005, US oil imports peaked at 2.52 billion barrels of oil. With the Great Recession of 2008, oil imports dropped 14 % to 2.48 billion barrels [9].

Many Americans may be under the impression that most of our oil is imported from the 12 OPEC nations,[9] with much of that oil coming from the Middle East.

However, Canada was our largest supplier of oil in 2013 (33.5 %), followed by Mexico at 9.8 % and the rest of Central and South America at 16.4 %, for 2013. Over 50 % of our oil imports came from the American continent. Only 39.7 % of our imports came from OPEC nations, as shown in Fig. 2.4 [10]. As shown in Fig. 2.5, since 2008, the percentage of oil imported from countries on the American continent has been steadily increasing compared to oil imported from OPEC countries, while the percentage imported from the Persian Gulf has been holding relatively steady near 20 %.

While US production is rising due to shale oil formations, the OPEC nations still have approximately 72.2 % of all proved oil reserves as summarized in Fig. 2.6, while North America has only 13.2 % of proved reserves; although unconventional oil exploration may help to alleviate this imbalance, OPEC will undoubtedly have a large influence on world oil supply for many decades. US oil production peaked in 1970 at 3.52 billion barrels and has been steadily declining

[8]Grid electricity currently costs less to power a battery electric vehicle in many parts of the nation now using fossil fuels, but if society imposed carbon standards and required the generation of clean electricity independent of fossil fuels, then electricity costs would be much higher; we show in Chap. 13 that hydrogen made from waste used in a fuel cell EV could be priced at a cost that is 10 % *less* than gasoline per mile driven in a regular car; but hydrogen may cost more than gasoline per mile initially until mass production of hydrogen fueling stations brings their prices down.

[9]OPEC = Organization of the Petroleum Exporting Countries which includes 12 nations (six in the Middle East—Iran, Iraq, Kuwait, Qatar, the United Arab Emirates, and Saudi Arabia); four from Africa (Libya, Algeria, Angola, and Nigeria), and two from South America (Ecuador and Venezuela).

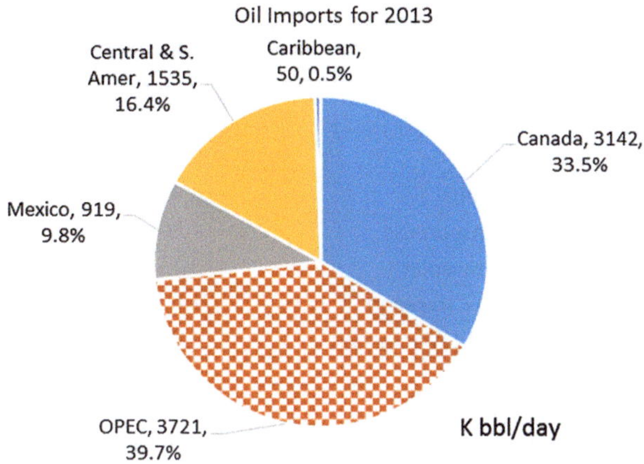

Fig. 2.4 US oil imports for 2013 in thousands of barrels per day

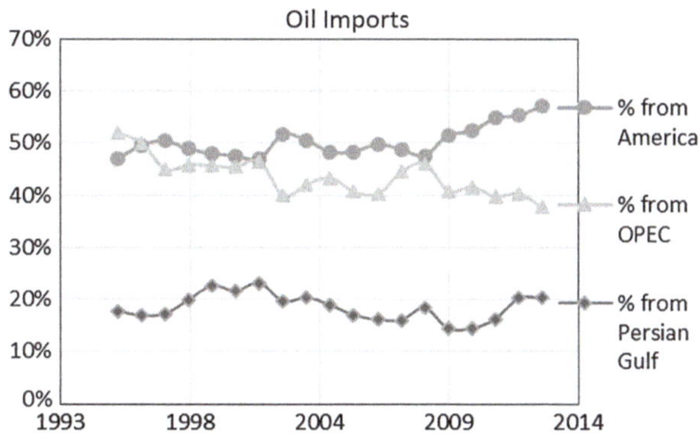

Fig. 2.5 Percentage sources of US oil imports over the last 8 years

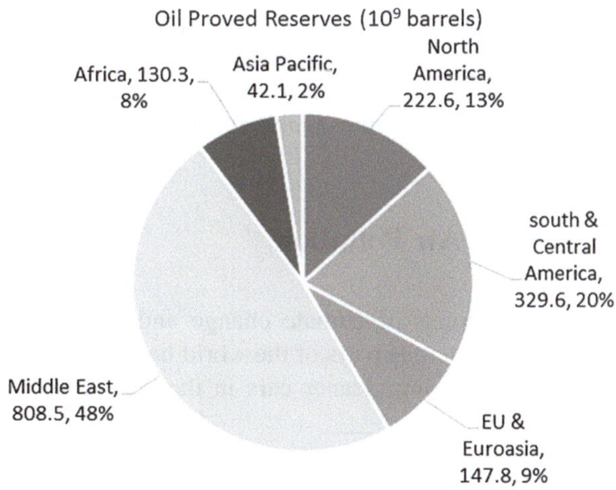

Fig. 2.6 Proved oil resources (trillion barrels) by region

Table 2.1 Oil available for LDV transportation under two oil import constraints for 2013: oil imported only from non-OPEC nations and oil only from the American continent

Import sources=>	Non-OPEC billion bbl/year	Americas only
2013 US production	2.72	2.72
Imports	2.24	2.05
Total oil available	4.9567	4.76252
Minus non-transport use	1.80	1.80
Available for transport	3.15	2.96
Minus non-LDV transport	1.87	1.87
Remainder for LDV transport	1.29	1.10

until hitting a low of 1.83 billion barrels in 2008; US oil production has rebounded since 2008 due to economic recovery and the discoveries of shale oil, reaching 2.72 billion barrels in 2013 [11].

The goals of importing all oil from "friendly" nations in the Americas[10] or only from non-OPEC nations are summarized in Table 2.1, based on 2013 data. In 2013, the USA imported 2.24 billion barrels from non-OPEC nations. Combining this with the US production of 2.72 billion bbl/year yields 4.96 billion bbls of oil available. Deducting the 1.8 billion bbl used for non-transportation needs leaves 3.15 billion bbl for all transportation. Subtracting off the 1.87 billion bbls used for other transportation needs (heavy trucks, trains, planes, etc.) leaves a residual of 1.29 billion bbl/year for LDVs. For the case of oil imported only from the American continent, the same tabulation indicates that only 1.1 billion bbl/year would be available to power LDVs.

We therefore set the goal of reducing LDV oil consumption to the range between 1.1 and 1.29 billion bbl/year, which would provide sufficient oil in some future energy crisis[11] without relying on OPEC or other nations outside the American continent. Note that these two goals are based on 2013 oil production, imports, and vehicle consumption. All three of these elements will undoubtedly change in the future, which will affect these goals. We assume that US oil production will rise, as will oil from other American and non-OPEC nations, which would increase the oil available for transportation.

2.3 Local Air Pollution

The twin threats of climate change and dwindling supplies of fossil fuels that reside in dangerous parts of the world have dominated the recent calls for developing and deploying cleaner cars in the developed nations, but local air pollution

[10]Excluding Venezuela, a founding member of OPEC.

[11]Even then, unless other nations of the world made similar reductions in their petroleum consumption, OPEC would still exert extraordinary influence in such a crisis.

remains a major threat to citizen's health in many parts of the world. Urban air pollution for some pollutants exceeded twice the World Health Organization limits in 2011 in seven "megacities" around the world[12] (a megacity is defined as having more than 10 million inhabitants). Mexico City was the most polluted city with more than twice the WHO-recommended safe levels of sulfur dioxide (SO_2), suspended particulate matter (SPM), carbon monoxide (CO), and ground-level ozone (O_3) [12].

Reducing urban air pollution in major cities around the world will require substantial reductions in motor vehicle emissions. Hence, the development and successful deployment of affordable clean vehicles in the USA will most likely be essential to reducing global urban air pollution.

Local air pollution is also still a major concern in some parts of the USA. Urban air pollution has been declining, but many areas of the USA still exceed the EPA limits for the six major "criteria pollutants"[13] covered by the Clean Air Act, as shown in Fig. 2.7. These criteria pollutants create significant health hazards for urban citizens and can also adversely impact crop growth in rural areas. Most of the non-attainment counties are in California, but criteria pollutants exceed the EPA standards in the Dallas/Fort Worth and Houston areas in Texas, and in many urban areas on the East Coast. Areas east of Cleveland and Pittsburg exceed the limits for 6 or 7 pollutants (red, pink or purple in Fig. 2.7).

While all-electric vehicles such as battery EVs and fuel cell EVs do not produce any significant criteria pollutants from the vehicles,[14] some pollutants are emitted in the process of producing the fuels for these vehicles (electricity and hydrogen).

The other major alternative vehicles such as hybrid electric vehicles and plug-in hybrids do have small gasoline- or diesel-powered internal combustion engines in the vehicles that emit significant criteria pollutants from their tailpipes, so it will be difficult to reduce air pollution if they continue to be part of the vehicle fleet.

The goal for the reduction of local air pollution should be to reach the EPA standards [13] for criteria pollutants in all regions of the country and for all cities of the world to achieve the World Health Organization standards for urban air pollution.

[12]Mexico City, Beijing, Cairo, Jakarta, Los Angeles, Sao Paulo, and Moscow [14].

[13]PM including two sizes: PM-2.5 and PM-10, referring to the size of the particles in microns, low-level ozone, CO, Sulfur oxides, Nitrogen oxides and lead.

[14]Actually, electric vehicles do generate particulate matter (PM) from brake linings during braking, and from tire wear on the roads. So the California "Zero Emission Vehicle" (ZEV) rating really only applies to tailpipe emissions and not to tire and brake wear.

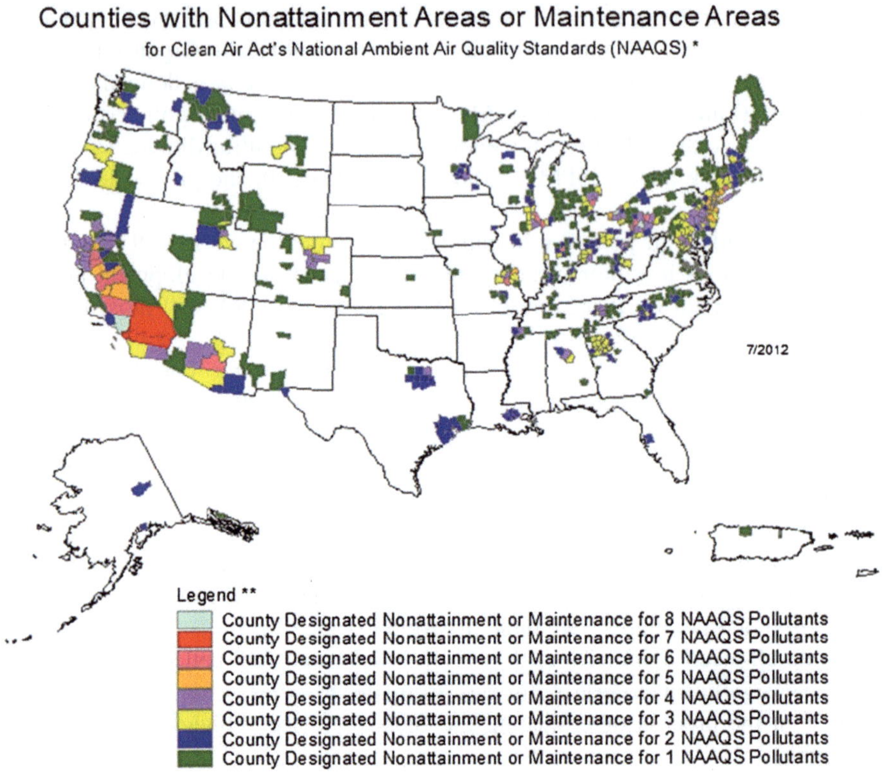

Fig. 2.7 Counties in the USA that did not meet the EPA standards for at least one urban air pollutant (*NAAQS* National Ambient Air Quality Standards for the six "criteria pollutants" specified in the Clean Air Act)

References

1. "Inventory of U.S. Greenhouse Gas Emissions and Sinks: 1990-2012, US Environmental Protection Agency Report EPA 430-R-14-003, April 15, 2014, available at: http://www.epa.gov/climatechange/Downloads/ghgemissions/US-GHG-Inventory-2014-Main-Text.pdf
2. D. Nakamura & S. Mufson, "U.S. and China announce climate deal," the Washington Post, November 12, 2014, page 1.
3. McKinsey & Company, "A Portfolio of power-trains for Europe: a fact-based analysis: The Role of Battery Electric Vehicles, Plug-in Hybrids and Fuel Cell Electric Vehicles" 2007, available at: http://cleancaroptions.com/html/a_portfolio_of_power_trains_for_europe_a_fact_based__analysis.pdf
4. "Energy Consumption by Sector and Source," U.S Department of Energy, Energy Information Administration, Available at: http://www.eia.gov/oiaf/aeo/tablebrowser/#release=AEO2014&subject=2-AEO2014&table=2-AEO2014&region=1-0&cases=ref2014-d102413a
5. Stacy Davis, Susan Diegel and Robert Boundy, "Transportation Energy Data Book, 32nd Edition, Oak Ridge National Laboratory report ORNL-6989, TABLE 2.6, July, 2013
6. J. M Diamond, *Guns, Germs, and Steel: The fates of human societies*, W.W. Norton & Company, NY, 1977. ISBN 0-393-31755-2

7. David Pimentel and Mario GIampietro *Food, Land, Population and the U.S. Economy*, Carrying Capacity Network, November 1994, available at: http://www.dieoff.com/page55.htm
8. *Danish Government to launch hydrogen Infrastructure Program & continue FCEV tax exemptions throughout 2015*, March 23, 2012, available at: http://www.h2euro.org/latest-news/hydrogen-hits-the-roads/danish-government-to-launch-hydrogen-infrastructure-program-continue-fcev-tax-exemptions-throughout-2015
9. Energy Information Agency (EIA), U.S. Department of Energy, "US Total Crude Oil & Product Imports," July 28, 2011, available at: http://www.eia.gov/dnav/pet/pet_move_impcus_a2_nus_ep00_im0_mbblpd_a.htm
10. Energy Information Administration, U.S. Department of Energy, "U.S. Imports by country of origin," available at: http://www.eia.gov/dnav/pet/pet_move_impcus_a2_nus_ep00_im0_mbblpd_a.htm
11. BP Statistical Review of World Energy, June, 2014, available at: http://www.bp.com/content/dam/bp/pdf/Energy-economics/statistical-review-2014/BP-statistical-review-of-world-energy-2014-full-report.pdf
12. Energy Information Administration, U.S. Department of Energy, available at: http://tonto.eia.gov/dnav/pet/hist/LeafHandler.ashx?n=PET&s=MCRFPUS1&f=M
13. National Ambient Air Quality Standards (NAAQS), U.S. Environmental Protection Agency, available at: http://www.epa.gov/airquality/greenbook/mapnmpoll.html
14. *Urban air pollution in world megacities*, available at: http://www.air-quality.org.uk/11.php

Chapter 3
Alternative Vehicle and Fuel Options

Abstract This chapter introduces the alternative vehicles and fuels analyzed in this book. We include five different electric vehicles and two conventional internal combustion engine (ICE) hybrid vehicles running on either natural gas or diesel fuel. We also analyze PHEVs running on cellulosic ethanol as a surrogate for other biofuels.

3.1 Electric Vehicles

We consider five different electric vehicle options.

3.1.1 Battery Electric Vehicles (BEVs)

When most people think of "electric vehicles," they envision battery electric vehicles that outnumbered gasoline ICVs in the early 1900s (more BEVs were sold in 1899 and 1900 than all gasoline and steam engine vehicles combined). These vehicles utilize electric motors to power the wheels, with all the electricity provided by batteries onboard the vehicle. Once these batteries run out of electricity, they must be recharged from some source, usually the electric power grid. In more modern times, GM produced and leased the two-seater EV-1 to help meet the California Zero Emission Vehicle (ZEV) mandate in the early 1990s. Consumers loved their EV-1s and were heartbroken when GM recalled these leased vehicles and crushed them as being too expensive to maintain.

More recently, Nissan introduced the Leaf in 2010, a full 4-passenger BEV with 80 miles range; Nissan sold approximately 76,891 Leafs through March 2015. Ten other OEMs[1] have sold a total of 139,153 BEVs by March 2015.

[1]BMW, Fiat, Ford, GM, Honda, Mercedes, Mitsubishi, Porsche, Smart, and Toyota.

C.E. (Sandy) Thomas, *Sustainable Transportation Options for the 21st Century and Beyond*, DOI 10.1007/978-3-319-16832-6_3

Table 3.1 Hybrid electric vehicles sold in the USA between 1999 and 2013

Toyota/Lexus	21,85,696	70.79 %
Honda/Acura	361,559	11.71 %
Ford/Mercury/LIncoln	292,171	9.46 %
Chevy/GMC/Buick/Cadilac	92,039	2.98 %
Hyundai	62,188	2.01 %
Nissan/Infiniti	39,474	1.28 %
Kia	24,003	0.78 %
Saturn	11,667	0.38 %
VW	6,575	0.21 %
Porsche	4,272	0.14 %
BMW	3,185	0.10 %
Mercedes-Benz	2,236	0.07 %
Mazda	1,144	0.04 %
Audi	1,124	0.04 %
Dodge/Chrysler	88	0.00 %
Totals	3,087,421	100 %

There are currently 11 BEV models available in the USA. Tesla Motors, a start-up company, has also produced approximately[2] 42,300 of their Model S BEV sedans through March 2015, for an industry total of 158,344 BEVs which represents approximately 0.06 % of the 250 million light-duty vehicles on the road in the USA.

3.1.2 Hybrid Electric Vehicles (HEVs)

HEVs were first introduced by Honda in 1999 with their Insight HEV although the Toyota Prius has since dominated the HEV market. An HEV, like a BEV, has an electric motor that can power the wheels, but it also has a small gasoline-powered internal combustion engine (ICE) that can provide power when the batteries run low or when extra acceleration is required. This ICE also typically can drive a generator to recharge the vehicle battery bank, which eliminates the "range anxiety" of the BEV. From 1999 through 2013, 3.09 million HEVs were sold in the USA as summarized in Table 3.1. Toyota accounted for 70.8 % of all HEVs sold (of which the Prius accounted for just under 50 % of total HEVs sold). These 3.09 million HEVs accounted for only 0.72 % of all 429 million LDVs sold in the USA over the period from 1999 through 2013. As of June 2014, 12 auto companies (Table 3.1) offer a total of 54 HEV models in the USA. HEVs sales totaled 3.08 million through 2014 [1].

[2]Tesla is the only auto-maker that does not state their vehicle sales monthly; analysts are also unsure of the number of Tesla BEVs delivered in the U.S. versus those going overseas so the number of Tesla BEVs in the US may be over-stated.

Table 3.2 Plug-in hybrid electric vehicle (PHEV) sales in USA through December 2014

	2010–2011	2012	2013	thru Dec '14	Totals
Chevy volt	7671	23,461	23,094	18,805	73,031
Toyota prius PHV			12,088	13,284	25,372
Honda PHEV			526	449	975
Ford C-max energi PHV			7154	8,433	15,587
Ford fusion energi PHV			6089	11,550	17,639
	7671	23,461	48,951	52,861	139,944

3.1.3 Plug-in Hybrid Electric Vehicles (PHEVs)

A plug-in HEV (PHEV) is an HEV that generally has a larger battery bank that can be plugged into a source of electricity to recharge the batteries while the vehicle is parked. This provides another layer of protection against running out of electricity, further dampening "range anxiety." General Motors produced the first commercial PHEV[3] which they call the "Volt" in 2010, although Ford built a prototype fuel cell PHEV in 2001.[4] Through December 2014, GM sold 73,031 Volt PHEVs, and three other auto companies are also selling four different PHEV models, with a total of 139,944 PHEVs sold in the USA by December 2014 (Table 3.2), which represents approximately 0.06 % of the 230 million LDVs on US roads [1].

The annual U.S. sales of these three EVs are shown in Fig. 3.1. HEVs have dominated EV sales since their introduction by Honda in 1999. The sales of BEVs and PHEVs are indistinguishable on this scale, so Fig. 3.2 shows an expanded scale of the same data to compare BEV and PHEV annual sales. The data for these two figures were taken from AFV sales through December of 2014. BEV sales were slightly higher than PHEV sales for 2014 as shown in Fig. 3.2.

3.1.4 Biofuel PHEVs

We also consider the use of biofuels, even though no automaker has yet to offer a PHEV fueled with biofuels; to maximize the benefits of biofuels, we assume that they are used in PHEVs that have higher efficiency than ICVs and therefore provide the maximum benefit of those biofuels in terms of reduced emissions and fossil fuel consumption. For this analysis, we use cellulosic ethanol as a surrogate for biofuels.

[3]However, GM calls their PHEV an "E-Rev" for extended range EV.

[4]The Ford PHEV was actually a fuel cell electric vehicle (FCEV) built on their Edge platform. Most FCEVs have a battery bank onboard to provide peak acceleration and to store energy during braking—called "regenerative braking" that improves vehicle efficiency by returning energy normally dissipated in the brake pads back into the battery to be used later; this Ford fuel cell PHEV was not offered for sale.

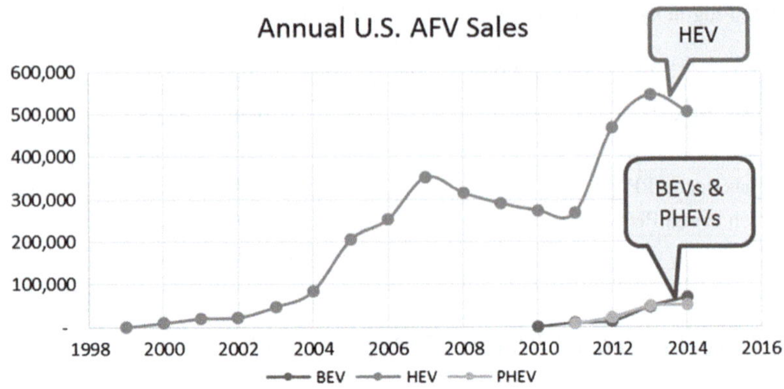

Fig. 3.1 Actual annual sales histories for alternative EVs in the USA

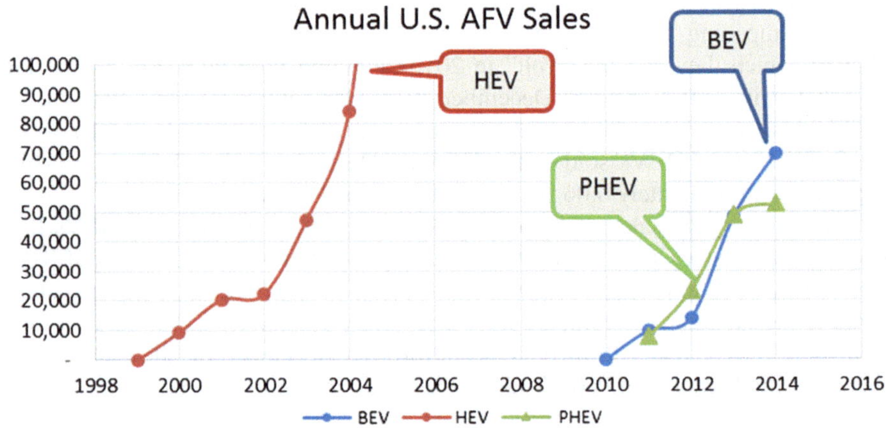

Fig. 3.2 Expanded scale for AFV sales to show the slight differences between BEV and PHEV sales since 2010

3.1.5 Hydrogen-Powered Fuel Cell Electric Vehicles (FCEVs)

A FCEV is fundamentally different from the other EVs discussed above, since the primary electricity source comes from a fuel cell on the vehicle and not from a battery.[5] The FCEV, like the BEV, is a pure electric vehicle and does not include an ICE such as those found on HEVs and PHEVs. A fuel cell is a device that produces electricity by an electrochemical reaction of hydrogen stored on the vehicle with oxygen from the air. As shown later in the book, the fuel cell system has

[5]However, as discussed above, most FCEVs also have a battery to provide peak acceleration and the ability to recover braking energy, called "regenerative braking."

lower mass, lower volume, and lower cost than a battery with similar energy and power rating. The hydrogen to power a FCEV is typically stored today as a high pressure gas[6] in high-strength carbon-fiber-wrapped tanks, similar to those used in millions of natural gas vehicles (NGVs)[7]around the world.

Honda began leasing five of their Clarity FCEVs in California in 2002. The upgraded 2005 version of the Honda Clarity had an EPA fuel economy of 72 miles/kg of hydrogen, which is equivalent to 74 miles/gal of gasoline, and a range of 190 miles on one tank of hydrogen. The National Renewable Energy Laboratory conducted an exhaustive on-the-road testing of 183 FCEVs produced by Daimler, GM, Hyundai-Kia and Ford in the 2003–2009 time period. These 183 FCEVs were driven a cumulative total of 3.07 million miles over this time period, with more than 25,000 pressurized hydrogen refueling events without any major incident [2]. National Laboratory engineers also tested two of the Toyota FCHV-adv. FCEVs, which are a fuel cell version of the Toyota Highlander SUV. These on-the-road tests in California showed that the Toyota FCEV SUV had a certified fuel economy of 68.3 miles/kg of hydrogen (equivalent to approximately 69.1 miles/gal of gasoline) and a range of 431 miles on one tank of hydrogen [3].

Several automakers have announced plans to begin producing FCEVs for sale in the 2015–2018 time period. Hyundai started FCEV production in 2014, and Toyota announced plans to begin producing FCEVs in the 2014–2015 time period. Mercedes, Ford, and Nissan announced a joint FCEV development plan in 2013 with a goal of starting commercial production of FCEVs by 2017 [4].

3.2 Other Alternative Fuel Vehicles

While the main premise of this book is that electric motors must replace the venerable ICE if we are to meet our national security and environmental goals, we also consider two alternative fuels, natural gas, and diesel fuel running in ICVs, since some commentators are suggesting that these fuels might be useful to meet some of our societal objectives. By analyzing these alternative fueled vehicles on the same basis as the EVs discussed above, the reader can judge the relative advantage of EVs over these non-gasoline ICVs. We assume that these fuels are used in HEVs to maximize their efficiency.

[6]BMW-tested vehicles with liquid hydrogen stored onboard, and in the 1990s, several auto companies attempted to build onboard reformers to convert liquid fuels such as methanol or even gasoline to hydrogen onboard the vehicle, but these reformers were either too expensive and/or did not survive the harsh vehicle environment.

[7]However, hydrogen is typically stored in modern FCEVs at higher pressure (70 MPa, 700-bar or 10,000 psi), compared to NGV tanks typically stored at lower pressures (25 MPa, 248-bar or 3600 psi).

3.2.1 Natural Gas Vehicles (NGVs)

NGVs were first introduced in large quantities in World War II when gasoline was in short supply in Europe. Many thousands of NGVs were built during and after the war, many of them do-it-yourself conversions of conventional gasoline ICVs.

3.2.2 Diesel-Fueled Internal Combustion Vehicles (ICVs)

We also compare the attributes of diesel-powered ICVs, since some commentators have noted that they help to reduce some pollution and reduce oil consumption due to their higher efficiency than conventional gasoline engines.

References

1. U.S. Alt Fuel Vehicle Sales, available at: http://green.autoblog.com/2014/06/04/may-2014-green-car-sales-numbers/
2. K. Wipke et al., *National fuel cell electric vehicle learning demonstration program*, The National Renewable Energy Laboratory, Report 3NREL/TP-5600-54860, July 2012; available at: http://www.nrel.gov/hydrogen/cdp_topic.html#infrastructure
3. K. Wipke at al., "Evaluation of range estimates for Toyota FCHV-adv. under open-road driving conditions," Savannah River National Laboratory and the National Renewable Energy Laboratory, Report # SRNS-STI-2009-00446, August 2009 available at: http://energy.gov/eere/fuelcells/downloads/evaluation-range-estimates-toyota-fchv-adv-under-open-road-driving
4. V. Vijayenthrian, "Mercedes, Ford and Nissan join forces on hydrogen fuel cell research, January 2, 3013. Available at: http://www.motorauthority.com/news/1081952_mercedes-ford-and-nissan-join-forces-on-hydrogen-fuel-cell-research

Chapter 4
Alternative Vehicle Market Potential

Abstract Some alternative vehicles have limited market potential due to factors such as short range, long fueling times, vehicle size limitations, and, for all vehicles, cost which is discussed in Chap. 5.

4.1 Range Limitations

As discussed in detail in Appendix A, battery electric vehicles (BEVs) have limited range, since both the weight and the volume of batteries become excessive at long ranges. As shown in Fig. 4.1, the mass of BEVs becomes excessive for ranges greater than approximately 500 km (311 miles), even if battery manufacturers are able to meet the US Advanced Battery Consortium (USABC) goals for battery-specific energy.[1] The FCEV, however, does not suffer from excessive mass at long ranges, since the extra hydrogen required for long range is very light, and the only increased mass is due to larger and heavier hydrogen tanks. At 500 km range, the FCEV weighs only 51 % of the BEV with batteries meeting the USABC goals. This extra mass of BEVs increases the mass, volume and cost of all vehicle components and reduces the net efficiency of the vehicle.

Batteries also require significant space in the vehicle, and this volume also grows with increased range as shown in Fig. 4.2. For the BEV, Fig. 4.2 shows just the volume of the battery system. For the FCEV, the volume shown includes the volume of the fuel cell system, the high-pressure hydrogen tanks, and a peak power battery bank used to augment power for acceleration and also to store the

[1]The model used to estimate the mass (Fig. 4.1) and volume (Fig. 4.2) was developed under contract to the Ford Motor Company, based on their lightweight Sable AIV (aluminum-intensive vehicle), part of DOE's "Partnership for a New Generation of Vehicles" (PNGV) under the Clinton administration in the 1990s. While further reduction in the glider mass and reductions in aerodynamic drag and rolling resistance could reduce the mass and volume of future BEVs, the relative relationship between the BEV and the FCEV should remain the same. Future reductions in vehicle body mass, aerodynamic drag, and rolling resistance will reduce mass and volume of both types of vehicles.

C.E. (Sandy) Thomas, *Sustainable Transportation Options for the 21st Century and Beyond*, DOI 10.1007/978-3-319-16832-6_4

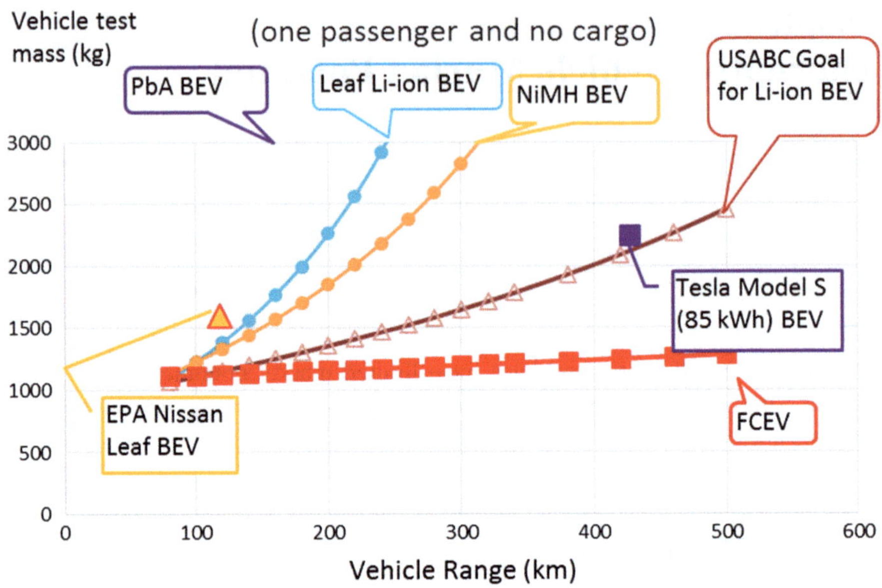

Fig. 4.1 Estimated mass of BEVs and FCEVs as a function of vehicle range assuming one passenger (the driver!) and no cargo (from Appendix A)

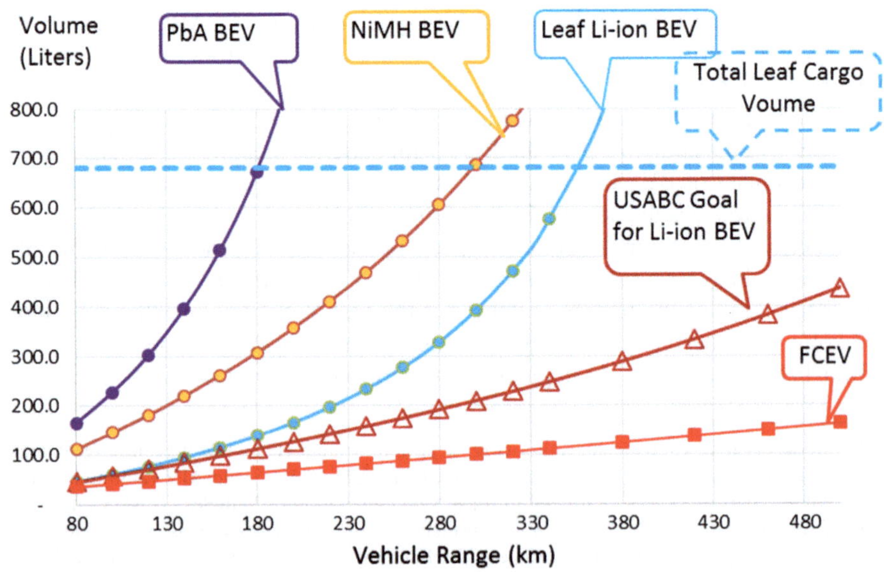

Fig. 4.2 Interior volume required to store the electricity generating components as a function of range

energy from regenerative braking (in other words, the volume of the fuel cell system that directly replaces the function of the battery bank—to produce electricity for the motor). At 500 km, the FCEV power train occupies only 40 % of the space occupied by a hypothetical BEV with batteries achieving the USABC commercialization goals.

4.2 Fueling Time Limitations

The time to refuel a vehicle is an important consideration for car buyers. The standard is the time it takes to refuel conventional gasoline vehicle of 3–5 min. Longer refueling times may limit market acceptance of an alternative vehicle.

Fuel cell electric vehicles (FCEVs) are fueled much like a conventional car. A high-pressure hose is connected to the fueling receptacle on the car, and hydrogen gas is forced in much the same as gasoline pouring into a conventional car tank and almost identical to natural gas filling an NGV. Filling times fall into the 3–5 min range in actual on road testing by National Laboratories.

BEVs take much longer to recharge a depleted battery.

Figure 4.3 shows the time required to recharge BEV batteries for two charger levels as a function of BEV range. As shown in Fig. 4.1, the mass of a BEV increases rapidly with added range. Much of this increased mass is due to a larger battery bank; more energy is then required to propel the heavier vehicle. The larger battery requires more charging energy which in turn requires more charging time.

In Fig. 4.3, the "Level 1" charging assumes a typical 120-V, 20-A home circuit; "Level 2" refers to a 240-V, 40-A circuit that might be used for high-power

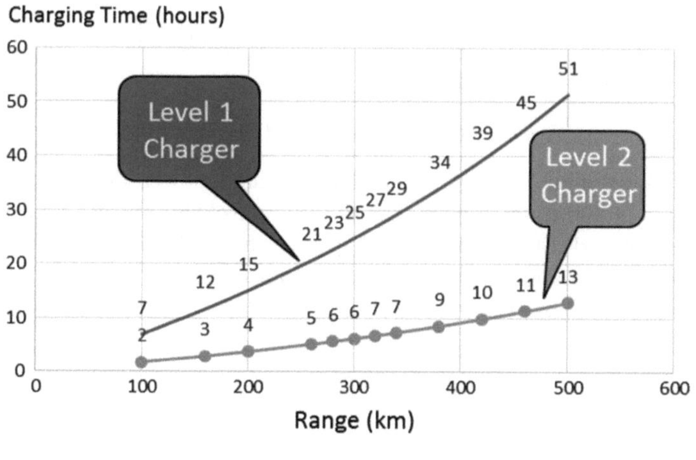

Fig. 4.3 Time to recharge BEVs versus vehicle range for Level 1 and Level 2 chargers (numbers on graph are charging time in hours) assuming USABC battery goals are met

appliances such as ranges or air conditioners. For a home Level 1 charger, a BEV with 500 km (311 miles) range would weigh 2,455 kg as shown in Fig. 4.1 and would require a battery bank storing 141.18 kWh. Assuming a maximum of 70 % depth of discharge to avoid battery damage, the charger must supply $141.18 * 0.7 = 98.8\,\text{kWh}$. The maximum power for a home 120-V, 20-A circuit is 1.92 kW,[2] which would require 51.5 h of charging to provide 98.8 kWh of energy:

$$98.8\,\text{kWh}/1.92\,\text{kW} = 51.5\,\text{h}.$$

A high-power 240-V, 40-A home circuit with a maximum power level of 7.68 kW would reduce the charging time for 500 km range to approximately 12.9 h or close to an overnight charge at home.

Higher power DC chargers are feasible, but costly, and not all batteries can accept very high power levels without a reduction in their lifetime or even destruction of the battery.

One metric to compare gasoline and hydrogen refueling with battery charging is to consider the *power* represented by gasoline or hydrogen surging through a hose. For example, pumping 15 gallons of gasoline into a car tank in 3 min represents a power transfer rate of more than **ten million watts** (10.1 MW) or 10 MW of power!

Similarly, from a detailed DOE testing assessment of 183 FCEVs that traveled over 3 million miles and were refueled a collective total of 25,811 times, the average hydrogen fueling rate was 0.77 kg/min [1]. Since 1 kg of hydrogen has an energy content of 33.3 kWh (LHV), then this flow rate of 0.77 kg/min is equivalent to a power draw of:

$$0.77 \times 33.31 \times 60 = 1539\,\text{kW} \tag{4.1}$$

or 1.54 **megawatts** (MW) of power.

The corresponding power flow for EV charging circuits is, for Level 1, 120 V, 20 A, and the maximum power draw (see footnote 2) is:

$$P_{\text{t1}} = 120 \times 20 \times 0.8/1000 = 1.92\,\text{kW} \tag{4.2}$$

And for Level 2 240-V, 40-A charging circuits,

$$P_{\text{L2}} = 240 \times 40 \times 0.8/1000 = 7.68\,\text{kW}. \tag{4.3}$$

So, as summarized in Table 4.1, the demonstrated power flow of hydrogen to refuel FCEVs is 800 times higher than the flow of electricity for a Level 1 home charging circuit and 200 times faster than a Level 2 240-V circuit. We conclude that it is much easier and faster to pass molecules of gasoline or hydrogen through a hose than to pass electrons through a wire.

[2]The National Electric Code requires circuits to be limited to 80 % of maximum power.

Table 4.1 Comparison of power flow rates for gasoline, hydrogen, and electricity

			Power (kW)	Ratio to level 1	Ratio to level 2
Gasoline	5	gallons/min	10,100	5260	1315
Hydrogen	0.77	kg/min	1539	802	200
BEV level 2 charger	240 V	40 A	7.68	4.00	1.0
BEV level 1 charger	120 V	20 A	1.92	1.00	0.25

4.3 Size Limitations

Due to the size and volume limitations of even advanced batteries as shown in Figs. 4.1 and 4.2, BEVs will most likely be limited to relatively small vehicles. After a very thorough evaluation of BEVs and FCEVs in Europe conducted by McKinsey & Company with confidential data from various automobile companies, McKinsey [2] concluded that:

> "BEVs are ideally suited to smaller cars and shorter trips, i.e. urban driving." And "FCEVs are the lowest carbon solution for medium/larger cars and longer trips. These car segments account for 50 % of all cars and 75 % of CO_2 emissions, hence replacing one ICE with one FCEV achieves a relatively high CO_2 reduction."

McKinsey concludes that BEVs could at best replace 50 % of all cars in Europe that emit only 25 % of all CO_2 emissions, so BEVs alone cannot relieve our climate change threats.

While this book deals almost exclusively with light-duty road vehicles, we should also note that BEVs cannot address pollution and oil consumption due to heavy-duty vehicles, buses, trucks, and even ships and airplanes, while hydrogen has already been tested as a fuel for aircraft and for heavy-duty trucks. For example, Vision Industries has demonstrated the use of fuel cells for heavy-duty Class 8 trucks used to haul freight around marine terminals [3]. Their Class 8 fuel cell truck has a curb weight of 7710 kg (17,000 lbs) and can haul trailers weighing 56,284 kg (80,000 lbs) with a 65-kW fuel cell system. The range of this truck on one tank of hydrogen is 124 km (200 miles). Vision Industries has also received a grant to develop a fuel cell refuse truck.

4.4 Market Surveys

Deloitte conducted a survey of potential BEV buyers in 2011. The survey revealed that 68 % of car buyers wanted a vehicle with at least 124 km (200 miles) range and 58 % expected refueling times of less than 2 h [4]. Neither expectation is likely to be met for most BEVS with the information reported above.

Another 2011 USA Today/Gallup poll found that 57 % would never buy a BEV no matter how high the price of gasoline [5].

4.5 BEV Market Potential

As analyzed in Appendix B, we estimate that only 31 % of light-duty vehicles could become BEVs due to limited range, long refueling times, and lack of fixed parking spaces as summarized in Table 4.2. This estimate is based in part on the Carnegie Mellon University estimate that only 56 % of all LDVs have access to an off-street parking place where a battery charger outlet could be installed [6].

Table 4.2 Estimated percentage of US light-duty vehicles that could become BEVs, assuming the Carnegie Mellon University estimate that only 56 % of all LDVs have access to off-street parking where a battery charging outlet could be installed

	Vehicles on the road (%)	2012 % sales	Average OTR and sales (%)	% that could be BEVs	% that have access to off-street parking	Net BEV percentages (%)
Small car	24.7	25.1	24.9	100	56	14.0
Midsize car	18.1	21.7	19.9	50	56	5.6
Large car	8.3	6.2	7.3	30	56	1.2
Small wagon	2.2	3.6	2.9	95	56	1.6
Midsize wagon	1.0	0.3	0.7	50	56	0.2
Large wagon	0.2	0.0	0.1	30	56	0.0
Small NT SUV	0.3	0.0	0.1	90	56	0.1
Midsize NT SUV	1.5	4.0	2.7	40	56	0.6
Large NT SUV	0.8	2.9	1.8	25	56	0.3
Small pickups	2.9	0.0	1.46	80	56	0.7
Midsize pickup	10.6	0.2	5.4	30	56	0.9
Large pickup	0.1	9.8	4.9	20	56	0.6
Small van	6.7	0.0	3.35	80	56	1.5
Midsize van	0.6	4.9	2.8	30	56	0.5
Large van	10.3	0.2	5.2	20	56	0.6
Small T SUV	1.1	0.0	0.5	80	56	0.2
Midsize T SUV	9.0	10.3	9.7	30	56	1.6
Large T SUV	7.4	10.6	9.0	20	56	1.0
All small vehicles	37.9	28.8	33.3	Total estimated BEV potential		31.1

References

1. K. Wipke et al., The National Fuel cell electric vehicle learning demonstration final report," the National Renewable Energy Laboratory Report NREL/TP 5600-5486, July, 2012. Available at: http://www.nrel.gov/hydrogen/proj_learning_demo.html
2. McKinsey and Company, "A portfolio of power-train options for Europe: a fact-based analysis, The role of battery electric vehicles, plug-in hybrids and fuel cell electric vehicles," available at: http://ec.europa.eu/research/fch/pdf/a_portfolio_of_power_trains_for_europe_a_fact_based__analysis.pdf
3. Vision Industries, "Heavy Duty Class A Trucks" available at: http://visionmotorcorp.com/tyrano.asp
4. E. Loveday, "EV Expectations out of touch with reality, August 1, 2011, available at: http://green.autoblog.com/2011/08/01/survey-electric-vehicle-expectations-out-of-touch-with-reality/
5. Z. McDonald, "Bias alert: survey says that 57 % would never buy an electric vehicle," available at: http://www.plugincars.com/survey-57-percent-would-never-consider-electric-vehicle-107237.html
6. E.J. Traut et al., "U.S. residential charging potential for electric vehicles," Carnegie Mellon University, *Transportation Research Part D* 25 (2013) 139-145. Available at: http://www.cmu.edu/me/ddl/publications/2013-TRD-Traut-etal-Residential-EV-Charging.pdf

Chapter 5
Alternative Vehicle Cost Estimates

Abstract Alternative vehicles should ideally cost no more than the conventional gasoline vehicles to speed market acceptance. While new technology vehicles will initially cost more, the goal should be to have equal or lower cost in the long term when these alternative vehicles are mass produced. This chapter summarizes a bottom-up assessment of alternative vehicle costs and also reviews two major studies of the likely mass production costs of alternative vehicles: one by Kromer and Heywood at MIT, and the other by the consulting firm McKinsey & Company.

5.1 Initial Capital Costs

5.1.1 Battery Electric Vehicles

5.1.1.1 Battery Costs

The battery pack dominates the cost of a BEV. For example, the Nissan Leaf BEV has a 24-kWh battery pack (of which 16 kWh is actually usable assuming a 67 % depth of discharge.[1])

The Argonne National Laboratory projected Li-ion battery pack prices at $706/kWh (baseline) to $250/kWh (optimistic) in 2000 [1]. A Duke University paper from 2009 projected that by 2020, Li-ion battery costs could range $300/kWh (optimistic) to $500/kWh (baseline) to $730/kWh (pessimistic) [2]. More recently, McKinsey & Company estimated current battery pack costs between $500/kWh and $600/kWh, with projected cost reductions to $200/kWh by 2020 and to $160/kWh by 2025 [3]. These battery cost estimates are summarized in Table 5.1.

[1]Batteries should not be completely discharged; the useful life of the battery decreases rapidly with increased depth of discharge.

C.E. (Sandy) Thomas, *Sustainable Transportation Options for the 21st Century and Beyond*, DOI 10.1007/978-3-319-16832-6_5

Table 5.1 Estimated costs of lithium battery packs in $/kWh

(All $/kWh)	ANL	Duke		McKinsey		
	2000	2009	2020	2012	2020	2025
Pessimistic		480	300			
Baseline	706	780	500	550	200	160
Optimistic	250	880	730			
USABC	150					

Table 5.2 Leaf 24-kWh battery pack costs based on projected future battery cost from Table 5.1

	ANL	Duke		McKinsey		
	2000	2009	2020	2012	2020	2025
Pessimistic	$–	$11,520	$7200			
Baseline	$16,944	$18,720	$12,000	$13,200	$4800	$3840
Optimistic	$6000	$21,120	$17,520			
USABC	$3600					

Table 5.3 Estimated costs for a Tesla model S 85-kWh battery system based on projected future battery costs from Table 5.1

	ANL	Duke		McKinsey		
	2000	2009	2020	2012	2020	2025
Pessimistic	$–	$40,800	$25,500			
Baseline	$60,010	$66,300	$42,500	$46,750	$17,000	$13,600
Optimistic	$21,250	$74,800	$62,050			
USABC	$12,750					

The impact of these battery costs on the total battery pack cost for the Nissan Leaf with a 24-kWh battery bank is shown in Tables 5.2 and 5.3 for the Tesla Model S with an 85-kWh battery pack. Thus, the Nissan Leaf battery pack would cost at least $3600 even if the USABC goal of $150/kWh could be reached and could still be as much as $12,000–$17,500 by 2020 based on the other battery cost projections cited above. The much larger Tesla battery pack would cost at least $12,750 and possibly as much as $42,000–$62,000 each by 2020 based on expert projections of future battery costs. These costs would be a major limitation for widespread use of BEVs for larger vehicles, both in terms of initial vehicle prices as well as for battery warranty costs.

5.1.2 Electric Motor/Controller Costs

The costs per kW for the motor and controller will be similar for any electric vehicle and will not significantly affect the relative costs of BEVs versus FCEVs (except that BEVs will generally weigh much more than FCEVs for a given

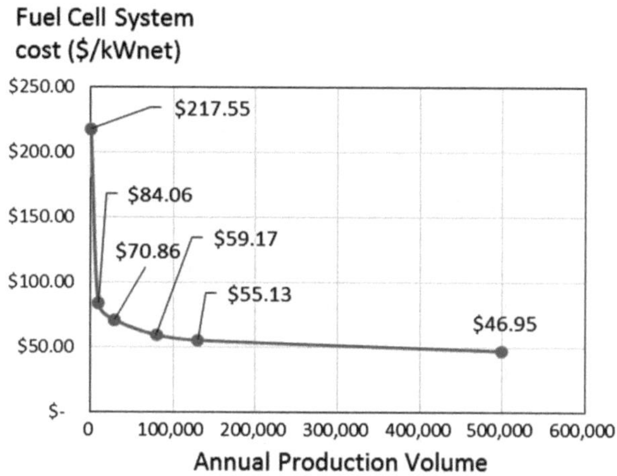

Fig. 5.1 Estimated mass production cost of automotive fuel cell systems

vehicle range and will therefore require larger and more expensive motors to provide adequate acceleration). The cost of motor/controllers is not heavily dependent on power; however, the cost of the motor/controller system will not have a large impact on the relative costs of BEVs versus FCEVs. For example, Kromer and Heywood at MIT [4] estimate that a motor controller will cost $200 plus $15/kW.

As shown in Fig. 5.1 in the previous chapter from data derived in Appendix A, a BEV will weigh much more than a FCEV due to mass compounding. To achieve a 500-km (311-mile) range, a BEV would weigh 2455 kg assuming that the USABC battery goals were met, while a FCEV with that range would weigh only 1283 kg. As a result, the BEV motor/controller power required to achieve a zero to 96.5 km/h (60 mph) speed in 10 s would be 181 kW, while the FCEV would only require a 95-kW motor/controller. The BEV motor/controller would therefore cost approximately $2915, while the FCEV motor/controller would cost $1625, using the Kromer and Heywood cost estimate from the previous paragraph. Although the BEV motor/controller costs 1.8 times more than the FCEV motor/controller, the motor/controller accounts for only 13.5 % of the total FCEV cost and 6.9 % of the total BEV cost as shown below, so the impact of the Motor/controller cost is relatively minor.

5.2 FCEV Cost

Three components are required on a FCEV to replace the function of the battery bank on a BEV:

- The fuel cell system;
- The hydrogen storage system; and
- A peak power battery for acceleration and to store regenerative braking energy.

5.2.1 Fuel Cell System Costs

Brian James of Strategic Analysis, Inc. [5] has been conducting detailed mass production cost estimation for the US Department of Energy for many years.[2] The estimated cost of automotive fuel cell systems in automotive scale production (500,000 units per year) has dropped significantly as the fuel cell systems have developed over the years from over $105/kW net estimated in 2006 to just over $46/kW net in 2012 for automotive fuel cell systems.

Furthermore, the estimated cost is relative relatively flat with production volume after a "knee" in the vicinity of 30,000 units per year, which is a quite modest production volume in automotive production terms. Kromer and Heywood of MIT use a similar FC system cost estimate of $50/kW for their base case. Thus, an 80-kW net automotive fuel cell system would cost approximately $3760–$4000 in automotive production volumes.

James points out that his fuel cell estimates are actual *costs* to the automobile companies and do not include markups for overhead or general and administrative (G&A) expenses, warranty, and profit that would be necessary to estimate *prices* of the fuel cell to consumers; we assume in this chapter that the other costs are also at the cost level and do not include markups to reflect prices,[3] so our comparisons here are valid.

5.3 Hydrogen Storage Costs

Kromer and Heywood of MIT [4] use a hydrogen storage tank cost estimate of $15/kWh of hydrogen or just under $500 per stored kg[4] of hydrogen at 350-bar pressure and $18/kWh or $600/kg for 700-bar tanks (which we assume here, since all major auto companies have now settled on using 700-bar pressure for hydrogen storage).

5.3.1 Peak Power Battery Costs

Kromer and Heywood used a peak power battery cost of $750/kWh for a FCEV, and $250/kWh for a high energy battery used in BEVs [4].

[2]This work originated at Directed Technologies, Inc. in the mid-1990s under contract to the Ford Motor Company, whose main costing analyst taught James and his team the Design for Manufacturing and Assembly (DFMA) methodology of Boothroyd Dewhurst, Inc. DTI was subsequently acquired by Strategic Analysis, Inc., where James and his team continue their detailed assessment of fuel cell costs for the Department of Energy.

[3]In Appendix E, we do include a markup to reflect *prices* for a stationary fuel cell system, since we are using prices for other equipment in that appendix.

[4]1 kg of hydrogen has an energy content of approximately 33.31 kWh (lower heating value).

5.4 Summary of EV "Bottom-Up" Cost Estimates

The estimated cost of BEVs and FCEVs versus vehicle range is shown in Fig. 5.2. For ranges less than 120 km (74.5 miles), the estimated BEV power train cost is less than the cost of a FCEV power train. But, the FCEV cost is much less than the estimated BEV cost for ranges greater than 120 km as the battery mass and cost dominates the BEV cost. A BEV battery bank designed for 500 km (311 mile) range would cost approximately 3.5 times more than the hydrogen-based power train for a FCEV with 500 km range.

The BEV power train cost has only two components: the battery bank and the motor/controller.[5]

The FCEV power train has five major components to replace the function of a battery (to provide electricity to the motor): the fuel cell system, a peak power battery, the hydrogen storage tank, and the motor/controller. As shown in Fig. 5.3, the hydrogen tank cost increases with range since larger tanks (more hydrogen) are required for longer range, while the other three components are fairly constant with range (although each component increases slightly due to mass compounding as the FCEV mass increases as shown in Table 5.4).

5.5 MIT EV Cost Estimates

Kromer and Heywood from MIT analyzed the projected cost of various electric vehicles in 2007 [4]. The results of their very detailed cost assessments for the 2030 time period are summarized in Fig. 5.4, which shows the incremental costs of various alternative vehicles relative to a conventional gasoline ICV. According to this MIT assessment, the FCEV will have a lower incremental cost than either a BEV, a PHEV-40, or a PHEV-30. Note that the BEV analyzed by MIT has a range of only 200 miles (483 km), while the FCEV has a range of 390 miles (628 km). In addition, the MIT team apparently did *not* consider the impact of mass compounding on these vehicles, so these differences would have been much larger, particularly for the BEV if they compared vehicles with long range plus mass compounding effects.

Here is how Kromer and Heywood of MIT evaluated the potential of BEVs:

> The (BEV) vehicle with a 200-mile range does not approach the utility of a conventional vehicle, although with a widespread rapid-charge infrastructure, it is conceivable that this reduced range would be acceptable to consumers. However, the battery cost of the 200-mile vehicle is *prohibitively expensive* without significant incentives or additional benefit. A vehicle with an electric range on the order of 400 miles could obviate the need for

---

[5]Actually, other components of the BEV will be heavier and costly more as the range increases. Thus, the BEV body, the suspension system, the brakes will all have to become larger and more costly as the BEV mass grows with range, but estimating the cost of these vehicle components affected by mass compounding was beyond the scope of this book.

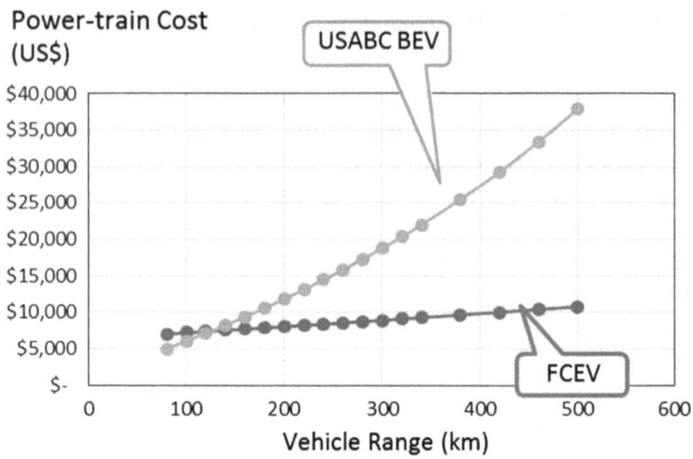

Fig. 5.2 Estimated power train costs for BEVs and FCEVs as a function of vehicle range, assuming that the USABC battery-specific energy and cost targets are met

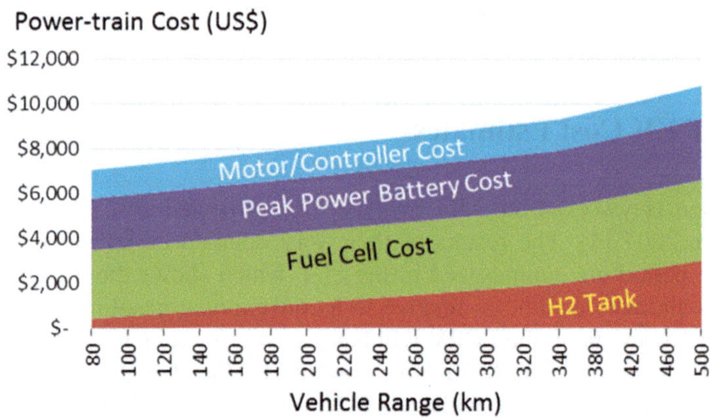

Fig. 5.3 Breakout of FCEV power train cost components versus range

extensive recharging infrastructure (beyond that installed in residences) as it approaches the distance that individual conceivably drive in a day, but this vehicle *is prohibitively expensive and heavy. The huge weight penalty of the battery pack* cancels much of the vehicle's energy efficiency benefit [4].

5.6 McKinsey & Company Cost Estimates

Another McKinsey report that analyzed the impact of alternative vehicles in the EU [6] includes estimates of both the initial cost of alternative vehicles and the total cost of ownership (TCO). The TCO includes the initial purchase price plus

Table 5.4 BEV and FCEV cost basis with USABC battery goals for the BEV as a function of vehicle range, including the impact of mass compounding

BEV costs							FCEV costs									
Range (km)	USABC BEV mass (kg)	Battery energy req'd (kWh)	Battery cost (US$)	Mtr/crtl Power req'd (kW)	Mtr/cntrlr cost (US$)	Total BEV power train cost (US$)	FCEV mass (kg)	Fuel cell power req'd (kW)	Fuel cell cost (US$)	Pk pwr battery energy (kWh)	Pk pwr battery cost (US$)	H_2 req'd (kg)	H_2 tank cost (US$)	Mtr/crtl power req'd (kW)	Motor/ control-ler cost (US$)	Total FCEV cost (US$)
80	1035	14.9	$3717	72.0	$1280	$4997	1063	64.6	$3037	3.05	$2288	0.75	$452	71.8	$1,276.8	$7053.3
100	1078	18.9	$4723	74.8	$1323	$6046	1071	65.1	$3060	3.08	$2308	0.95	$567	72.3	$1,285.1	$7220.1
120	1124	23.1	$5763	77.8	$1367	$7130	1080	65.6	$3083	3.11	$2329	1.14	$682	72.9	$1293.4	$7388.2
140	1170	27.4	$6838	80.8	$1413	$8251	1089	66.1	$3107	3.13	$2350	1.33	$799	73.5	$1301.8	$7557.3
160	1219	31.8	$7951	84.0	$1460	$9412	1097	66.6	$3131	3.16	$2370	1.53	$916	74.0	$1310.2	$7727.7
180	1269	36.4	$9104	87.3	$1509	$10,613	1106	67.1	$3155	3.19	$2391	1.72	$1034	74.6	$1318.7	$7899.3
200	1320	41.2	$10,298	90.7	$1560	$11,858	1115	67.6	$3179	3.22	$2413	1.92	$1153	75.2	$1,327.3	$8072.0
220	1374	46.1	$11,536	94.2	$1613	$13,148	1124	68.2	$3203	3.25	$2434	2.12	$1273	75.7	$1,335.9	$8,246.0
240	1430	51.3	$12,820	97.8	$1667	$14,487	1133	68.7	$3228	3.27	$2455	2.32	$1394	76.3	$1,344.6	$8,421.1
260	1488	56.6	$14,153	101.6	$1724	$15,877	1142	69.2	$3252	3.30	$2477	2.53	$1515	76.9	$1,353.3	$8,597.6
280	1548	62.2	$15,538	105.5	$1783	$17,321	1151	69.7	$3277	3.33	$2498	2.73	$1637	77.5	$1,362.1	$8,775.2
300	1610	67.9	$16,978	109.6	$1844	$18,822	1160	70.3	$3302	3.36	$2520	2.94	$1761	78.1	$1,371.0	$8,954.2
320	1675	73.9	$18,476	113.9	$1908	$20,384	1169	70.8	$3327	3.39	$2542	3.14	$1885	78.7	$1,379.9	$9,134.4
340	1743	80.1	$20,036	118.3	$1974	$22,011	1179	71.3	$3353	3.42	$2564	3.35	$2010	79.3	$1,388.9	$9,315.9
380	1887	93.4	$23,358	127.7	$2116	$25,473	1197	72.4	$3404	3.48	$2609	3.77	$2262	80.5	$1,407.1	$9,682.8
420	2044	107.9	$26,978	138.0	$2270	$29,248	1216	73.5	$3456	3.54	$2655	4.20	$2519	81.7	$1,425.6	$10,055.1
460	2216	123.8	$30,940	149.2	$2438	$33,378	1236	74.7	$3509	3.60	$2701	4.63	$2779	83.0	$1,444.3	$10,432.8
500	2405	141.2	$35,294	161.6	$2624	$37,917	1255	75.8	$3562	3.66	$2748	5.07	$3043	84.2	$1,463.3	$10,816.1

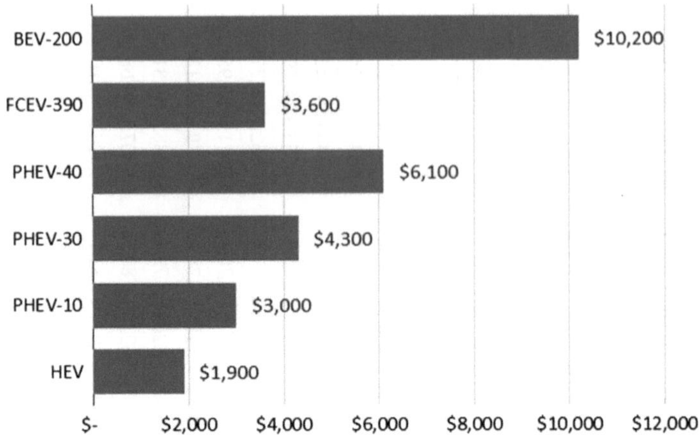

Fig. 5.4 Estimated incremental cost of alternative vehicles relative to a conventional gasoline internal combustion vehicle in the 2030 time period (the numbers after a vehicle name indicate the all-electric range of those vehicles in miles)

maintenance and fuel costs, including the cost of the fueling infrastructure per vehicle. This McKinsey & Company report is particularly authoritative, since their detailed assessment included the participation of 11 automotive companies,[6] 5 oil and gas companies, 2 utilities, 3 industrial gas companies, 4 electrolyzer companies, and 5 other organizations. Of particular interest to the estimation of alternative vehicle costs, the automobile companies provided proprietary vehicle cost information to the McKinsey analysts that was handled in "clean rooms" so that no auto company had access to their competitors' confidential cost information. The McKinsey report states that:

> BEVs are ideally suited to smaller cars and shorter trips, i.e. urban driving (including new transportation models such as car sharing).

Regarding FCEVs, McKinsey & Company goes on to say:

> With a driving range and performance comparable to ICEs, FCEVs are the lowest carbon solution for medium/larger cars and long trips. These car segments account for 50 % of all cars and 75 % of CO_2 emissions, hence replacing one ICE with one FCEV achieves a relatively high CO_2 reduction.

The largest car set is called the "J-segment" in the EU, represented by SUVs including the Hyundai Tucson, the Toyota RAV2, Toyota Highlander, and Ford Explorer. The estimated vehicle prices and the TCO for alternative vehicles in 2050 are shown in Fig. 5.5 for the SUV "J" Class. They are projecting that a large FCEV will cost slightly less than a BEV by 2050 (€30,400 vs. €33,300); the FCEV also has lower TCO than a BEV (€41,400 vs. €44,300). However, the McKinsey report does not specify the ranges of these vehicles, so it is difficult to

[6]BMW, Daimler, Ford, GM, Honda, Hyundai, Kia, Nissan, Renault, Toyota, and Volkswagen.

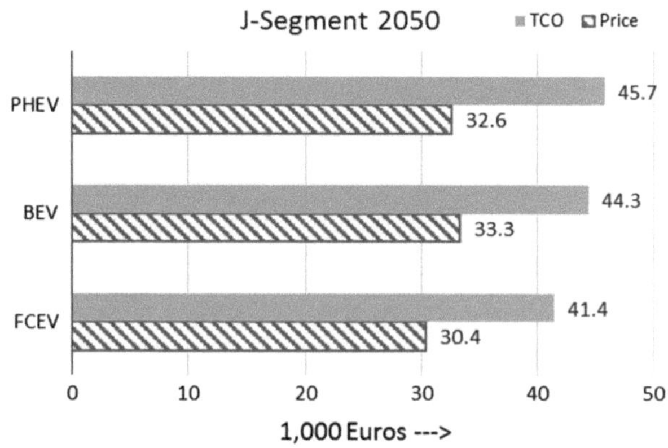

Fig. 5.5 Estimated price and total cost of ownership (TCO) for BEVs and FCEVs in 2050 for the "J-segment" SUVs

compare their results with the MIT data or our own calculations reported above as a function of vehicle range. In addition, the McKinsey report did not mention mass compounding, which further reduces the difference between BEV and FCEV costs that they calculated.

References

1. L. Gaines & R. Cuenca, "Costs of Li-Ion Batteries for Vehicles," Report # ANL/ESD-42, Argonne National Laboratory, May 2000.
2. D. Anderson, "An Evaluation of current and future cost for Lithium-Ion batteries for use in electrified vehicle powertrains," Nicholas School of the Environment, Duke University, May 2009.
3. R. Hensley, J. Newman & M. Rogers, "Battery Technology Charges Ahead," McKinsey & Company, July, 2012, available at: http://www.mckinsey.com/insights/energy_resources_materials/battery_technology_charges_ahead
4. M. Kromer & J. Heywood, "Electric power train opportunities and challenges for the U.S. light-duty vehicle fleet," Sloan Automotive Laboratory, Laboratory for Energy and Environment, Massachusetts Institute of Technology, Cambridge, Massachusetts, Report # LFEE 2007-03XP, May 2007, available at: http://web.mit.edu/sloan-auto-lab/research/beforeh2/files/kromer_electric_powertrains.pdf
5. B. D. James, "Fuel Cell Transportation Cost Analysis, Strategic Analysis, Inc., available at: http://www.sainc.com/service/SA%202011%20Update%20of%20Cost%20Analysis%20of%20Light%20Duty%20Automobile%20Fuel%20Cell%20Power%20Systems.pdf
6. McKinsey & Company, "A Portfolio of Power-Train options for Europe: a fact-based analysis: the role of battery electric vehicles, plug-in hybrids, and fuel cell electric vehicles," undated. Available at: http://ec.europa.eu/research/fch/pdf/a_portfolio_of_power_trains_for_europe_a_fact_based__analysis.pdf

Chapter 6
Fuel Infrastructure Cost

Abstract All electric vehicles will need some type of fuel infrastructure other than the ubiquitous gasoline and diesel fuel stations. Many observers note the high cost of installing sufficient hydrogen stations to fuel FCEVs, but BEVs and PHEVs will also require new charging stations to replenish depleted batteries. While a hydrogen station may cost much more than a battery charging station, one hydrogen station may support hundreds or even thousands of FCEVs since each FCEV can be refueled in a few minutes, while one charging outlet can support at most one or two BEVs per day due to their long charging times. As a result, we show in this chapter that the *lowest* estimated cost of charging outlets per BEV is *5 times* the *highest* cost estimate of hydrogen fueling cost per FCEV.

6.1 Hydrogen Fueling Stations

The cost of a hydrogen fueling station can be in the one million dollar range. But one station may serve hundreds or thousands of FCEVs, so the cost per vehicle is manageable. Initial hydrogen fueling stations produced hydrogen on-site either by electrolyzing water or by steam reforming of natural gas to make hydrogen. Other stations utilized hydrogen delivered by pipeline, much like the ubiquitous natural gas pipelines in the USA, or trucked-in hydrogen in either gaseous or liquid form. More recently, many warehouses have begun converting their fork lift trucks and other material handling equipment to run on hydrogen-powered fuel cells in place of batteries. Most of these fork lift truck applications have relied on trucked-in liquid hydrogen in the USA as the lowest cost option. The estimated cost of liquid hydrogen (LH2) fueling stations is summarized in Table 6.1 based on data from Yang and Ogden [1] and data from a Nextant report [2] for the hydrogen vaporizer. The estimated cost varies from $560 K to over $2.5 million depending on station capacity.

To determine how many FCEVs can be supported by one hydrogen station, consider the average annual miles traveled in the USA and the FCEV fuel economy. For 2013, the average USA light-duty vehicle traveled 11,876 miles

C.E. (Sandy) Thomas, *Sustainable Transportation Options for the 21st Century and Beyond*, DOI 10.1007/978-3-319-16832-6_6

Table 6.1 Estimated installed costs for LH2 fueling stations

LH2 station capacity (kg/day)	LH2 Storage	LH2 Pump	Vaporizer	GH2 Cascade	Dispensers (1 to 3)	8 % Site Prep & permits	8 % markup	10 % Eng. & design	5 % contingency	30 % installation	Total installed cost	# of FCEVs supported with CF = 96 %	Cost per FCEV
150	$36,016	$18,802	$13,649	$20,374	$230,667	$25,561	$27,605	$37,267	20,497	$129,132	$559,570	416	$1,346
300	$65,080	$37,005	$24,910	$40,749	$230,667	$31,873	$34,423	$46,471	25,559	$161,020	$697,755	832	$839
400	$83,457	$49,054	$32,418	$54,332	$230,667	$35,994	$38,874	$52,479	28,864	$181,841	$787,979	1,109	$711
1000	$186,538	$120,733	$77,461	$135,829	$461,333	$78,552	$84,836	$114,528	62,990	$396,840	$1,719,640	2,772	$620
1500	$267,804	$180,066	$114,997	$203,743	$692,000	$116,689	$126,024	$170,132	93,573	$589,509	$2,554,538	4,158	$614

Table 6.2 Average daily hydrogen consumption for FCEVs

VMT/year	11,876
FCEV fuel economy	93.9 miles/kg
Hydrogen per day	0.35 kg/day

Table 6.3 Average number of FCEVs supported by fueling stations and average infrastructure cost per FCEV

LH2 station capacity (kg/day)	Total installed cost	# of FCEVs supported with CF = 96 %	Cost per FCEV
150	$559,570	416	$1,346
300	$697,755	832	$839
400	$787,979	1,109	$711
1000	$1,719,640	2,772	$620
1500	$2,554,538	4,158	$614

according to the Oak Ridge National Laboratory [3]. The measured fuel economy of the Toyota Highlander SUV FCEV was 68.3 miles per kg of hydrogen consumed; however, the new Toyota Mirai passenger vehicle FCEV is smaller, with lower cross section and lower aerodynamic drage and a reported fuel economy of 93.9 miles/kg (a reported range of 650 km on 4.3 kg of hydrogen [4]. As shown in Table 6.2, the average daily hydrogen consumption for a FCEV would then be 0.35 kg/day.

With an average consumption of 0.35 kg/day, a fueling station with a capacity of 150 kg/day could therefore support up to 150/0.35 = 433 FCEVs, but this assumes a station capacity factor of 100 %. Assuming an average station capacity factor of 96 %, then this 150 kg/day station could support 416 FCEVs[1] as summarized in Table 6.3. The total infrastructure cost varies between $614/FCEV for very large stations to as much as $1,346/FCEV for a small 150 kg/day station. The average fueling station cost per vehicle is approximately $965/FCEV for the small stations (150, 300, and 400 kg/day) that would be utilized for early FCEV deployments or an average of $826 per FCEV including all capacity stations from 150 kg/day to 1,500 kg/day.

6.2 Battery Charging Stations

Several organizations have estimated the cost of providing electrical charging outlets for BEVs and PHEVs. The industry has established three primary charging levels for batteries:

[1]Note that this is **not** 416 FCEVs fueled *per day*, but the number of FCEVs that could be *supported* by one station with this capacity. Thus, if the FCEV was refueled once every 7 or 8 days, this would then imply that only 25–29 FCEVs would be refueled per day, which is quite modest, given the proven refueling times of 3–5 min.

Table 6.4 Nominal ranges and charging times for BEVs

			Charging time (hours)		
	Nominal AER range (miles)	EPA rated range (miles)	Level I 120 V	Level II 240 V	Level III 480 + volt
Nissan Leaf BEV	100	73	16	8	80 % in 1/2 h
Chevy Volt PHEV	40	?	10	4	
Ford Focus BEV	≈100	?	18–20	3–4	??
Mitsubishi MiEV	83		22.5	6	80 % in 1/2 h
Toyota RAV4 BEV	83.5	103	44	6	

1. Level 1, the customary US home circuit of 120 V and 20 A, with a maximum power of 1.92 kW.[2]
2. Level 2, 240 V and 40 A that might be used for electric ranges or air conditioners or other high power appliances with a maximum power of 7.68 kW.
3. Level 3, fast charger, 480 V, 3 phase with powers in the range from 60 kW to as much as 150 kW.

One of the most credible battery charging cost estimates is from Dr. Jan Kreider of Boulder, Colorado, a professional engineer who solicited bids from three electrical contractors to retrofit an existing 300-car parking garage to add charging outlets to each parking space [4]. The design called for 10 % Level 3 fast chargers and 90 % Level 2 240-V outlets. The average cost of the three competitive bids was $12,400 for each Level 2 outlet and $106,000 for each Level 3 fast charger.

One key parameter is the number of BEVs or PHEVs that could be supported by one outlet, which depends on the charging time required. As shown in Table 6.4, most BEVs require 4–8 h to charge with a Level 2 240-V outlet, and even the Chevy Volt PHEV requires 4 h at 240 V. So most Level 2 outlets will only support one BEV each day, unless there is extraordinary coordination between car owners, such as at a workplace where one BEV is charged in the morning and another in the afternoon from the same outlet. Thus, the charging infrastructure cost for Level 2 240-V chargers varies between $6200 and $12,400 per BEV based on the Kreider & Associates bids for retrofitting an existing parking garage in Boulder, Colorado, depending on whether two BEVs could share the same outlet; more than two PHEVs could potentially share an outlet if they only needed an hour or two to top off their batteries.

[2]The US National Electric Code limits circuits to 80 % of their maximum power, so $120 \times 20 \times 80\,\%/1000 = 1.92$ kW for a Type 1 circuit.

Table 6.5 Estimated component costs for various battery charging circuits by RMI

	Level 2 home		Level 2 parking garage		Level 2 curbside		Level 3 fast charger	
	Lo	Hi	Lo	Hi	Lo	Hi	Lo	Hi
Hardware	$450	$1000	$1500	$2500	$1500	$3000	$12,000	$35,000
Electrician	$100	$350	$1240	$2940	$800	$1500	$1600	$3000
Elect. materials	$50	$150	$210	$510	$150	$300	$300	$600
Other materials			$50	$100	$50	$150	$100	$400
Other labor			$250	$750	$2500	$7500	$5000	$15,000
Mobilization	$50	$200	$250	$500	$250	$500	$600	$1200
Permits	$–	$100	$50	$200	$50	$200	$50	$200
Total	$650	$1800	$3550	$7500	$5300	$13,150	$19,650	$ 55,400
Average		$1225		$5525		$9225		$37,525

With a fast charger that could provide 80 % of the battery charge in half an hour, one could imagine 3 or 4 BEVs being supported by one fast charge outlet (assuming cooperation between the BEV owners or a third party who would sequence the BEVs through the charging routine), but this would still imply a cost between $26,500 and $35,300 per BEV.

The Idaho National Laboratory has also tabulated the costs for battery chargers [5]. They estimated that the cost for a Level 1 home outlet would be $878, and a Level 2 residential outlet would cost $2146. The Rocky Mountain Institute (RMI) [6] broke down the costs of various charging outlets as summarized in Table 6.5.

Melaina and Penev at NREL [7] have also estimated the costs for electrical outlets for commercial locations[3] as summarized in the "NREL average" row of Table 6.6, which also summarizes the average charging outlet costs including the average estimate for each of the three charging levels. The Level 1 charger is not adequate for BEVs, since it would take 10–44 h to recharge a BEV battery bank. We conclude that the cost of charging infrastructure for BEVs varies between $7252 per BEV for a Level 2 outlet and $14,971 per BEV for a fast charger, assuming that 4 BEVs could be sequenced into the charger each day. This compares with the estimated hydrogen fueling station cost of $614 per FCEV to at most $1,346/FCEV as shown in Fig. 6.1. Thus, the *low* estimate for commercial BEV charging infrastructure ($7252/BEV) costs 5.4 times more than the *high* FCEV estimate per vehicle ($1,346/FCEV).

[3]Melaina & Penev do estimate that a Level 2 residential outlet would cost an average of only $2318/BEV, and Table 6.6 represents the estimated cost for commercial or public charging outlets.

Table 6.6 Average public charging station costs and average cost per BEV

	Level 1	Level 2	Level 3
	1.92 kW	7.68 kW	60–150 kW
Idaho National Laboratory	$878	$2146	
Kreider & Associates		$12,400	$106,000
RMI Average		$7375	$37,525
NREL Average	777	$7085	$36,125
Average	$828	$7252	$59,883
# of BEVs	1	1	4
$/BEV	$828	$7252	$14,971

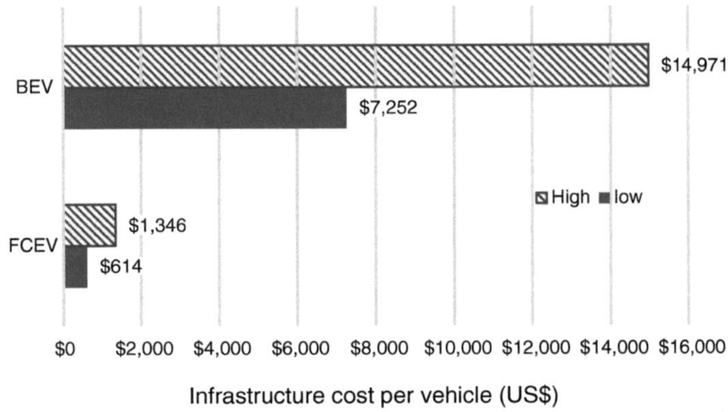

Fig. 6.1 Comparison of the low and high estimates for commercial BEV charging infrastructure per BEV and the hydrogen infrastructure cost per FCEV

6.3 McKinsey & Company Infrastructure Cost Estimates

McKinsey & Company analyzed a long 40-year rollout of BEVs and FCEVs in Europe [8]. They estimated that it would cost €101 billion over 40 years to install an EU-wide hydrogen infrastructure. They also estimated that installing a charging infrastructure for BEVs and PHEVs would cost five times more at €540 billion over 40 years.

On a per vehicle basis, McKinsey analysts assumed 100 million FCEVs over this 40-year period and 200 million BEVs and PHEVs,[4] so the infrastructure costs per vehicle are estimated at €1010 per FCEV and approximately €2700 per BEV or PHEV.

[4]Although there are twice as many BEVs and PHEVs as FCEVs in the McKinsey assessment, the BEVs in particular are smaller cars traveling shorter distances, since the larger vehicles replaced by FCEVs account for 75 % of all EU light-duty vehicle GHGs.

References

1. C. Yang & J. Ogden, Institute for Transportation Studies, U of California at Davis, "Hydrogen Delivery and Refueling Infrastructure: Advanced Energy Pathways (AEP) Project" for the California Energy Commission, June 2007.
2. Nextant, "H2A Hydrogen Delivery Infrastructure Analysis Model and Conventional Pathway Option Analysis Results; Interim Report," DE-F636-05G15032, May 2008.
3. S. Davis et al., "Transportation Energy Data Book: Edition 32," Oak Ridge National Laboratory, ORNL-6989, July 2013. Available at: http://cta.ornl.gov/data/index.shtml
4. Dr. Jan F. Kreider, Kreider & Associates, Boulder, Colorado, first presented to the *Toyota Sustainable Mobility Seminar*, La Jolla, California, April 6, 2011, and published later with P. Curtis, "Developments in light duty vehicle life cycle analysis with applications to electric vehicles," Proceedings of the ASME 2011 5th International Conference on Energy Sustainability, Washington, D.C. August 7–10, 2011, available at: http://www.fuelsandenergy.com/papers/ES2011-54957.pdf
5. K. Morrow, D. Karner, & J. Francfort, "The U.S. Department of Energy Vehicle Technologies Program-Advanced Vehicle Testing Activities- Plug-in Electric Vehicle Charging Infrastructure Review," Battelle Energy Alliance Contract No. 5/8517, Report # INL/EXT-08-15058, November 2008. Available at: http://avt.inel.gov/pdf/phev/phevInfrastructureReport08.pdf
6. J. Agenbroad & B. Holland, "RMI: What's the true cost of EV charging stations?" May 8, 2014, available at: http://www.greenbiz.com/blog/2014/05/07/rmi-whats-true-cost-ev-charging-stations
7. M. Melaina & M. Penev, "Comparing the Infrastructure Costs of Hydrogen and Electricity", The DOE Fuel Cell Technologies Annual Merit Review, Arlington, Virginia, May 15, 2012.
8. McKinsey & Company, "A Portfolio of Power-Train options for Europe: a fact-based analysis: the role of battery electric vehicles, plug-in hybrids, and fuel cell electric vehicles," undated. Available at: http://ec.europa.eu/research/fch/pdf/a_portfolio_of_power_trains_for_europe_a_fact_based__analysis.pdf

Chapter 7
Computer Simulation Model Scenarios

Abstract A computer simulation model was written to analyze and compare the likely impacts on greenhouse gas (GHG) emissions and on fossil fuel consumption of converting from conventional gasoline-powered internal combustion engine vehicles (ICVs) to seven different alternative vehicle/fuel combinations including five electric vehicle combinations described in this chapter; these five scenarios will be used to estimate potential (GHG) and petroleum consumption in Chaps. 8 and 9: (1) Hybrid electric vehicle (HEV) scenario; (2) Gasoline plug-in hybrid electric vehicle (PHEV) scenario; (3) Biofuel PHEV scenario; (4) Battery electric vehicle (BEV) scenario; and (5) Fuel cell electric vehicle (FCEV) scenario. And two other non-conventional alternative fuel vehicles are analyzed in Chap. 11: (1) Diesel fuel HEV scenario; and (2) Natural gas HEV scenario.

7.1 Hybrid Electric Vehicle (HEV) Scenario

The Hybrid electric vehicle (HEV) scenario began in 1999 when Honda sold 17 of its small Insight HEVs. Sales of HEVs ramped up quickly when Toyota introduced over 5000 of their Prius HEVs in 2000. Total sales of HEVs reached nearly half a million in 2013, or approximately 3.4 % of all cars sold that year (but only 0.3 % of all LDVs on the road). As described in more detail in Appendix C, we matched the initial sales of HEVs to a logistics function over the twenty-first century, such that the sales of HEV ramps up rapidly as shown in Fig. 7.1. The resulting percentage sales of gasoline-powered HEVs and conventional gasoline internal combustion engine vehicles (ICVs) is shown over time in Fig. 7.2 for the HEV scenario.

C.E. (Sandy) Thomas, *Sustainable Transportation Options for the 21st Century and Beyond*, DOI 10.1007/978-3-319-16832-6_7

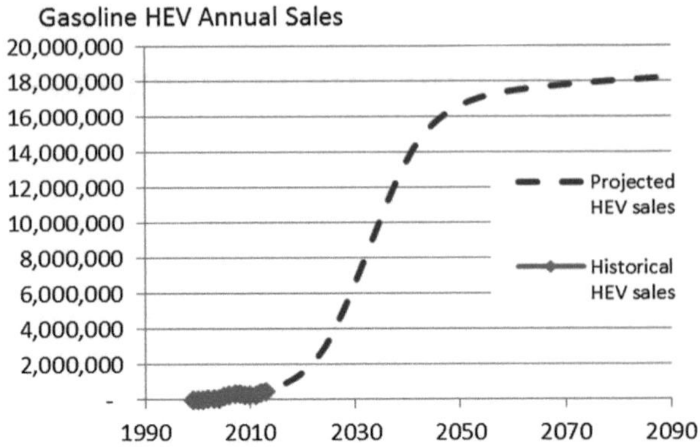

Fig. 7.1 Historic and projected HEV sales over the twenty-first century

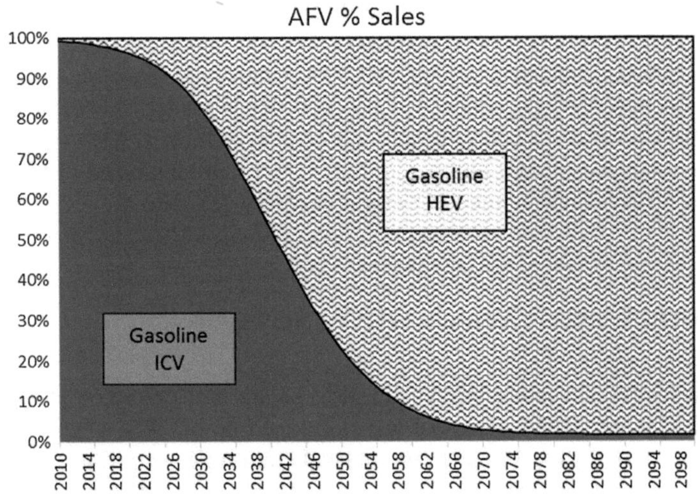

Fig. 7.2 Annual vehicle sales over the twenty-first century in the HEV scenario

7.2 Gasoline Plug-in Hybrid Electric Vehicle (PHEV) Scenario

The first commercial plug-in hybrid electric vehicle (PHEV), the Chevy Volt, was sold in 2011. We assume that the sales of PHEVs follow a similar trajectory to the HEV, but delayed in time (see Appendix C for details).

One limitation of PHEVs may be the lack of a place to plug in the vehicle at home. Researchers at Carnegie Mellon University have estimated that only 56 % of US motor vehicles have access to a dedicated parking space suitable for overnight charging, and that only 22 % of all vehicles have access to a charging outlet [1]. This will be a significant deterrent for battery EV (BEV) drivers, since

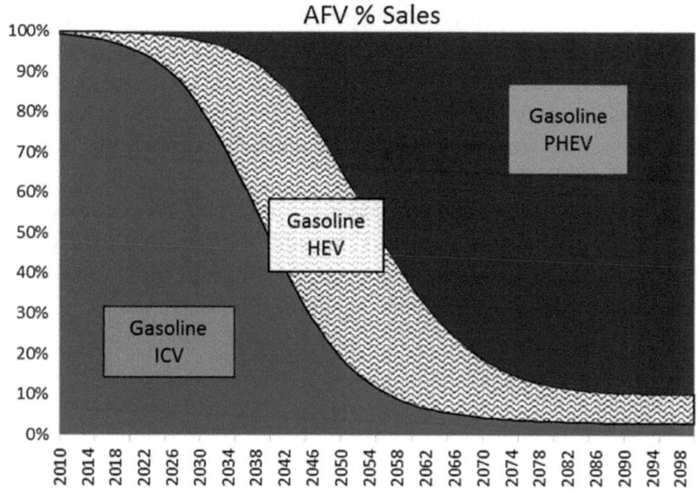

Fig. 7.3 Distribution of new car sales for the gasoline PHEV scenario

overnight charging will be nearly essential for a reasonable driving experience (assuming that charging at work would not be adequate for most BEV drivers due to the long charging times required). Lack of home charging will be less crucial for PHEV drivers since (a) they can rely on the gasoline engine to provide most range and (b) charging at work will be more suitable, since shorter times will be required to top off a PHEV battery than to charge a much larger BEV battery bank. Rather than using the 56 % estimate from Carnegie Mellon for those with access to off-street parking, we set a 75 % cap on PHEV sales, assuming that some PHEV drivers will be satisfied with charging their vehicles at work even though they do not have access to off-street parking at home.

The sales fractions for the gasoline PHEV scenario are summarized in Fig. 7.3. We have essentially superimposed PHEV sales on the HEV scenario, so the vehicles that are not PHEVs are made up of some ICVs and some HEVs. If all non-PHEV vehicles were conventional gasoline ICVs, then the scenario results would be less attractive. This also mimics the actual real-world conditions: HEVs like the Prius are here to stay and will be a part of any future automobile mix.

7.3 Biofuel PHEV scenario

The biofuel PHEV scenario is similar to the gasoline PHEV scenario, except that gasoline is gradually replaced with cellulosic ethanol as an example of biofuels.[1] The fraction of PHEVs fueled by ethanol is phased in gradually over the century

[1]Other biofuels could be used, but we have detailed GHG emissions data on ethanol. We assume that the ethanol is made from cellulosic feedstocks [such as corn stalks ("stover")] and not from corn to avoid the food versus fuel dilemma.

as shown in Fig. 7.4. The source of the ethanol is modified gradually as shown in Appendix C, starting with the current practice of making all ethanol from corn, transitioning over the century to ethanol made from corn stover (essentially the non-food stalk and roots from the corn plant), followed by ethanol made from switchgrass. Thus, the source of ethanol becomes "greener" and non-competitive with food supplies over the century.

The distribution of new car percentage sales over the century is shown in Fig. 7.5 for the biofuel PHEV scenario. As with the gasoline PHEV scenario, the biofuel PHEVs are superimposed on the previous mix of other alternative vehicles (including gasoline-powered PHEVs).

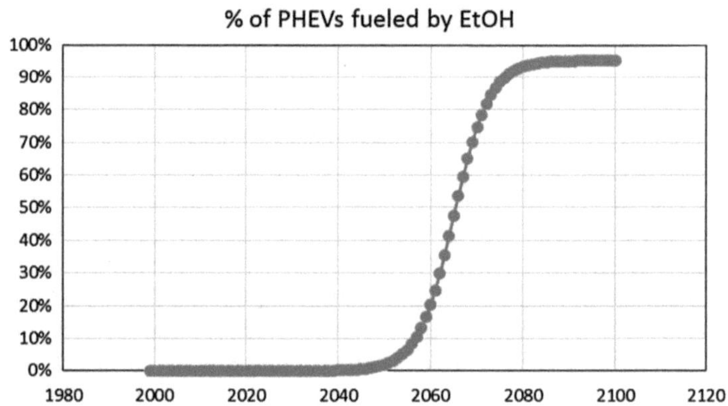

Fig. 7.4 Percentage of PHEVs fueled by ethanol

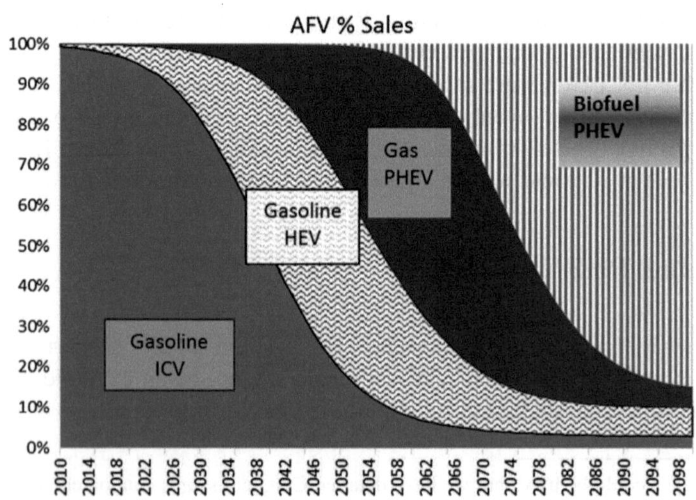

Fig. 7.5 Distribution of alternative vehicle sales over the century for the biofuel PHEV scenario

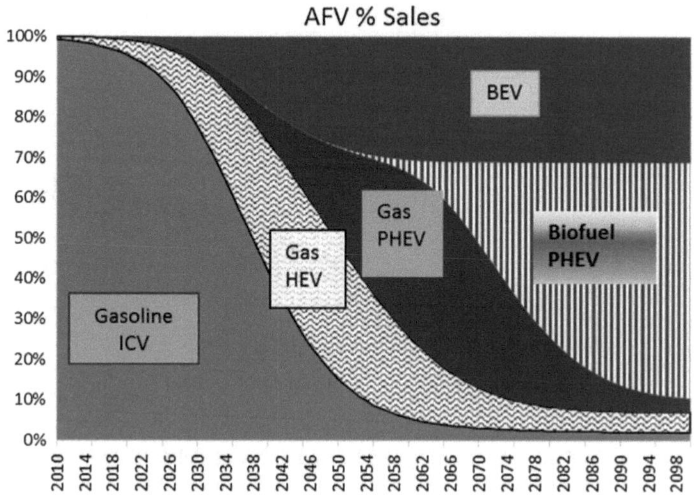

Fig. 7.6 Distribution of AFV sales over the century for the BEV scenario

7.4 Battery Electric Vehicle (BEV) Scenario

The BEV scenario is dominated by the probable number of BEVs that could be sold into the light-duty vehicle fleet as described in more detail in Appendix B, where the impact of the limited range of BEVs and the limited number of LDVs that have access to assigned off-street parking where they could be charged at night is used to estimate the maximum probable market penetration for BEVs in the USA. For the base case when Carnegie Mellon University research [1] showed that only 56 % of LDVs have access to assigned off-street parking, Appendix B shows that only 31 % of LDVs could likely be replaced by BEVs.[2] The details of BEV market penetration are summarized in Appendix C.

As with the other scenarios, we assume that the BEV sales are superimposed on the other AFV sales to minimize GHGs and oil consumption for the BEV scenario (compared to assuming that all non-BEV sales were conventional ICVs or even HEVs). The resulting distribution of new car sales percentages for the BEV scenario is shown in Fig. 7.6. By the end of the century, the bulk of the new car sales are biofuel PHEVs, the lowest carbon alternative. In effect, the BEV scenario displaces 31 % of the gasoline and biofuel PHEVS from the biofuel PHEV scenario. This scenario generates much lower GHGs and oil consumption than if we assumed that all non-BEV vehicles were gasoline ICVs or HEVs or even gasoline PHEVs.

In addition to introducing EVs vehicles over time, we also assume that the electricity used to charge vehicle batteries becomes "greener" over time, thereby

[2]In addition to BEV sales limited by off-street parking access, they would be limited by the ability to power larger, heavier vehicles.

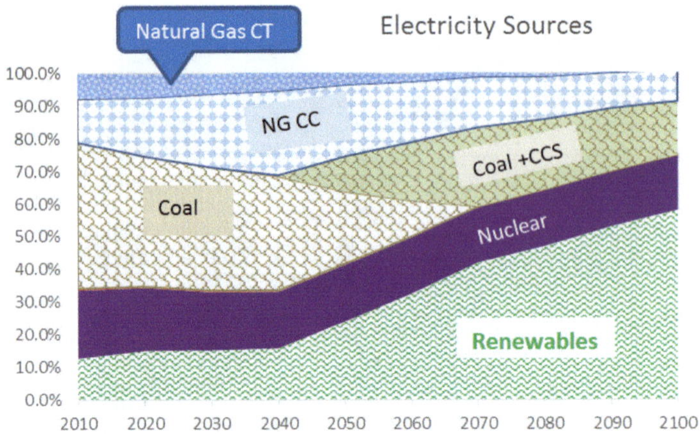

Fig. 7.7 "Greening" of the US electrical grid assumed in model (*CT* combustion turbine; *CC* combined cycle turbine; and *CCS* carbon capture and sequestration/storage.)

gradually reducing their GHG emissions. Coal-fired electrical generators are gradually replaced by alternative generators as shown in Fig. 7.7 including:

- Renewable energy electricity sources (solar, wind, wave, biomass, hydro-electricity, etc.)
- High-efficiency coal-based-integrated gasification combined cycle (IGCC) with carbon capture and sequestration (CCS), gradually replacing conventional coal power plants, with all conventional coal plants replaced by 2070.
- Nuclear power.

7.5 Fuel Cell Electric Vehicle (FCEV) Scenario

For the FCEV scenario, we assume no limits on FCEV sales, since FCEVs have the same range, fueling time, and power capacity as conventional ICVs, and fuel cells can be used to power heavy vehicles. The sales of FCEVs in the early years will be driven primarily by the California Zero Emission Vehicle (ZEV) mandate that requires auto makers to sell either BEVs or FCEVs over the next decade or so. The ZEV mandate should stimulate the sale of 20 to 30 thousand FCEVs per year initially in California. Seven other states[3] have "opted-into" the ZEV program and will account for another few thousand FCEV sales per year. Cumulative sales of ZEVs are anticipated to reach 3.3 million vehicles by 2025.

[3]Connecticut, Maryland, Massachusetts, New York, Oregon, Rhode Island, and Vermont.

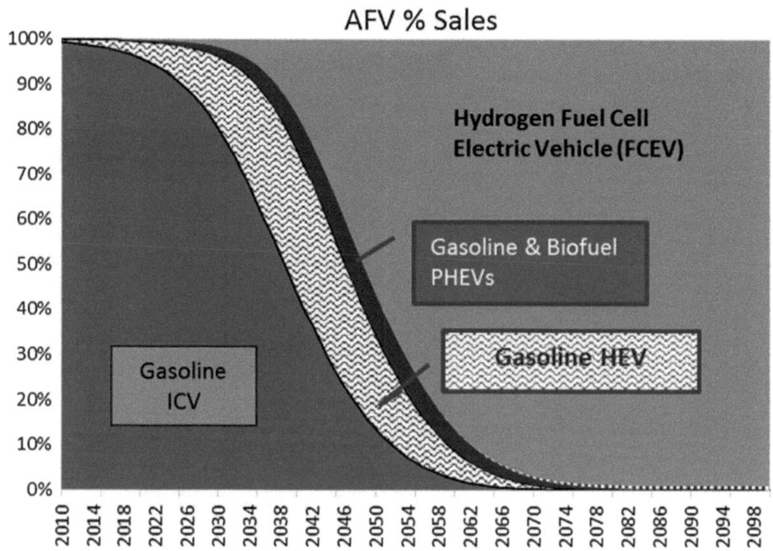

Fig. 7.8 Distribution of new car sales for the FCEV scenario

Appendix C summarizes the details of the assumed FCEV market penetration over time. Figure 7.8 shows the distribution of AFV sales over the twenty-first century for the FCEV scenario. As with the other scenarios, we assume that the FCEVs are superimposed on previous sales of other alternative fuel vehicles.

As with the BEV scenario and the "Greening of the grid," we assume that the production of hydrogen becomes greener over time. The model assumes that all hydrogen is made from natural gas initially, the lowest cost option today, but gradually transitioning to hydrogen made from biofuels such as cellulosic ethanol at the forecourt, followed by hydrogen made from a high-efficiency coal-integrated gasification combined cycle (IGCC) plant with (CCS) to prevent the CO_2 from entering the atmosphere plus hydrogen from biomass gasification with CCS and central electrolysis to produce hydrogen using either nuclear power or renewables.[4]

In addition, we consider the use of hydrogen made from wastewater treatment plants in Chap. 13 to provide an economic approach to installing initial hydrogen

[4]Not shown on this chart is hydrogen made from municipal solid waste and landfill gas, which would have lower net GHG emissions than hydrogen made by electrolysis from biomass since the conversion of waste at wastewater treatment plants reduces electricity-generated GHG emissions and also reduces the consumption of natural gas to heat the anaerobic digester tanks, further reducing the GHG emissions.

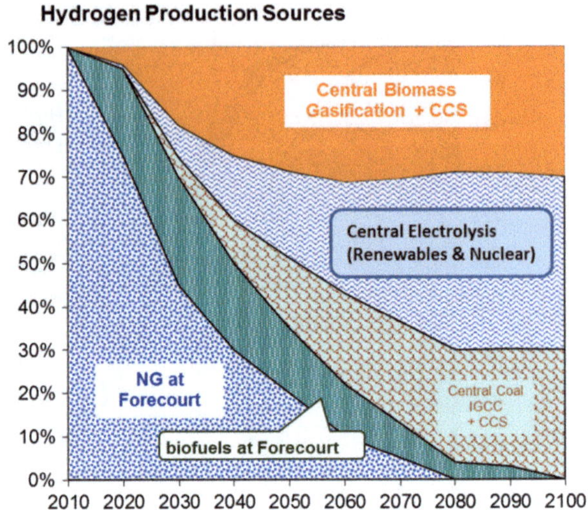

Fig. 7.9 Hydrogen production over the century, showing the gradual "greening" of hydrogen sources

stations with zero or negative carbon emissions; we have not included hydrogen from waste in Fig. 7.9 since the quantity of municipal solid waste is limited to less than 3 % of the FCEV market.

Reference

1. E.J. Traut et al., "U.S. residential charging potential for electric vehicles," Carnegie Mellon University, *Transportation Research Part D* 25 (2013) 139-145. Available at: http://www.cmu.edu/me/ddl/publications/2013-TRD-Traut-etal-Residential-EV-Charging.pdf.

Chapter 8
Greenhouse Gas Emissions for Alternative Vehicles

Abstract This chapter summarizes the key greenhouse gas (GHG) findings from this book. Our analysis shows that only the hydrogen-powered fuel cell electric vehicle (FCEV) can reach the goal of reducing GHGs by 80 % below 1990 levels. All other alternative vehicle options fall short of this goal.

We analyze the GHGs for all US alternative vehicles and fuels for two time periods:

- A near-term (2015–2020) "Business-as-usual" condition assuming no significant efforts to curb GHG[1] and single vehicle choices (either all-battery EV, plug-in hybrids, or all-fuel cell EV that provides an estimate of the maximum possible reductions in GHGs and oil consumption) and
- A longer term assessment assuming that significant climate change legislation provides incentives to reduce GHG emissions in both electrical generation and fuel production. In this case, we assume a set of multivehicle scenarios to capture the likely real-world situation where more than one vehicle type will be in the US light-duty vehicle (LDV) fleet, similar to the current on-the-road mixture with a combination of conventional gasoline ICVs and gasoline hybrid EVs.

In both time periods, we compare the GHG reductions of alternative vehicles and fuels with gasoline HEVs, which we assume will dominate the US car fleet by 2020 as car makers strive to meet the new corporate average fuel economy (CAFE) standards adopted in 2012. LDVs will have to meet an average fuel economy standard of 34.1 mpg (0.069 L/km) by 2016 and 54.5 mpg (0.043 L/km) by 2025, which could only be met by hybridizing most vehicles in the fleet, since there will probably be too few of the new alternative vehicles and fuels on the road by 2020 to impact the fleet average fuel economy for any automaker.

---

[1]With the election of many climate change "deniers" to the US Congress in 2010, chances for any climate change legislation in the USA seem to remote baring any major electoral shifts in the next few years, unless states such as California are able to launch a shift toward greener electricity through their proposed cap and trade legislation.

C.E. (Sandy) Thomas, *Sustainable Transportation Options for the 21st Century and Beyond*, DOI 10.1007/978-3-319-16832-6_8

In Chap. 2, we established a goal of reducing greenhouse gas (GHG) emissions from motor vehicles by at least 80 % and preferably 95 % below 1990 levels in order to avoid the worst aspects of global warming and climate change. Greenhouse gases [primarily carbon dioxide (CO_2), methane (CH_4), and nitrous oxide (N_2O)] accumulate in the atmosphere over many years. These gases seem to have an almost magical split personality: They are transparent to visible radiation, but opaque to thermal infrared radiation. As a result, during the day, they are transparent to the sun's visible radiation, allowing the sun to warm the Earth; we can see the results of solar heating during the day, but the human eye cannot observe what happens at night that is equally important for determining the Earth's average temperature.

The human eye can see only a very small fraction of electromagnetic radiation: We can see light varying from blue to red, but our world is full of many other wavelengths of radiation, including those below red (infrared) that we cannot see. We can feel infrared radiation after the flames die out at a campfire: Our face feels the heat due to infrared radiation from the hot coals. All objects with a temperature above absolute zero emit infrared radiation, including our own bodies. Our bodies are normally at 37 °C (98.6 °F), so we are frequently hotter than most objects in our surroundings. This is the basis of the military's night vision goggles: They include infrared cameras that display a picture of the infrared radiation in front of the soldier, effectively giving the soldier infrared vision; enemy soldiers literally "glow in the dark" in the views of these goggles.

Returning to the Earth's diurnal temperature balance, during the day greenhouse gases allow the sun's visible radiation to warm the Earth; at night, "greenhouse gases" reverse their function and act as a blanket over the Earth, blocking infrared radiation from leaving the Earth and thereby increasing the Earth's average temperature. This is the same dual nature of glass in a greenhouse: Glass allows in sunlight during the day so plants can grow via photosynthesis, but glass blocks thermal radiation from escaping at night thereby keeping the greenhouse warm, hence the origin of the term "greenhouse" gas. You can observe the effects of infrared thermal radiation on a cold winter night: In general, the coldest nights occur with a clear sky; nights are not quite as cold on winter nights with a cloud cover, since the water vapor in the clouds blocks the infrared radiation from cooling the Earth by transferring heat to the cold outer space.

Of course, we cannot see greenhouse gases, which are all invisible. And even if we could see these gases, we cannot sense the very small changes in the average Earth's temperature of only 0.6 °C (1.1 °F). The larger temperature changes that form the basis of our concerns about drastic climate change are predicted by complex computer models.

Some greenhouse gases are essential for our existence on Earth. Without any GHGs in the atmosphere, the Earth's temperature would drop as more thermal heat escaped into outer space, leading to a new ice age. But too much GHG has the opposite effect, and most scientists agree that temperatures will rise steadily over the twenty-first century as more GHGs accumulate in the atmosphere. Carbon dioxide used to account for approximately 280 parts per million (ppm) by volume (or 0.280 %) in the Earth's atmosphere; in other words for every million

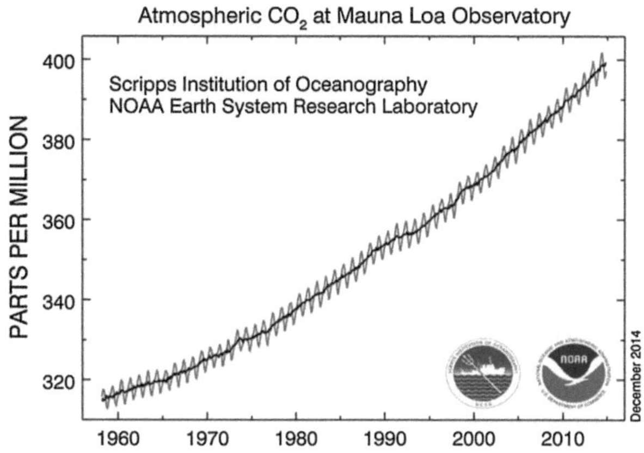

Fig. 8.1 Concentration of CO_2 gas in the atmosphere above Hawaii

gas molecules in the atmosphere, approximately 280 were CO_2 molecules. This 280 ppm value held relatively steady for approximately 800 years prior to 1800. But since the beginning of the industrial age, CO_2 content in the atmosphere has increased steadily up to 370 ppm in 2000 and to just over 400 ppm by July 2014 as shown in Fig. 8.1 [1]. This increase is attributed primarily to the burning of fossil fuels that always generate carbon dioxide. CO_2 is also absorbed by growing plants, since plants absorb CO_2 and release oxygen in the process of photosynthesis, while we humans do the reverse (breathing in oxygen and exhaling CO_2). As a result, CO_2 levels vary over the growing seasons by 3–9 ppm; in the Northern Hemisphere, CO_2 concentrations routinely drop in the spring, reaching a minimum in May each year as growing plants absorb CO_2 from the atmosphere, and rise again in the fall, reaching a peak in October when plants and trees become dormant or decay (which releases methane and CO_2 that are both potent greenhouse gases) and cannot absorb CO_2 to counteract the release of CO_2 from burning fossil fuels. The Northern Hemisphere dominates this seasonal fluctuation, since more plants and trees grow in the north than in the south. Another human activity, deforestation, also contributes to increased greenhouse gases in the atmosphere: As more trees are cut down to make room for additional agriculture in regions such as the Amazon, there are fewer plants to absorb CO_2, so atmospheric CO_2 increases.

While most climate scientists around the world agree that the Earth's temperature is rising and is most likely caused by human activities, there is a small but vocal group of scientists that either deny that there is any warming (although very few seem to cling to this position lately, since satellite measurements have verified that the average global temperature has already risen by 0.6 °C since the late nineteenth century), or they believe that, if there is warming, it is caused by natural phenomenon such as variable solar output or volcanic eruptions and not by human activities.

There does remain some uncertainty regarding the *consequences* of global warming, with different models predicting different results. Most scientists seem to agree that if the intergovernmental panel on climate change (IPCC) estimates of an average temperature rise of 1.4 to as much as 5.8 °C occur by 2100, then there may be very grave consequences, including sea level rise that could inundate many island nations and metropolitan cities, more violent storms, and a general decline in agriculture in the current agriculture regions of the world (or at least a steady migration of agricultural areas to the north in the Northern Hemisphere).

Here is what the Committee from the US National Academy of Sciences said in 2001 about climate change and the uncertainties involved [2]:

> "Greenhouse gases are accumulating in the Earth's atmosphere as a result of human activities, causing surface air temperatures and subsurface ocean temperatures to rise. Temperatures are, in fact, rising. The changes observed over the last several decades are likely mostly due to human activities, but we cannot rule out that some significant part of these changes is also a reflection of natural variability. Human-induced warming and associated sea level rises are expected to continue through the 21st century. Secondary effects are suggested by computer model simulations and basic physical reasoning. These include increases in rainfall rates and increased susceptibility of semi-arid regions to drought. The impacts of these changes will be critically dependent on the magnitude of the warming and the rate with which it occurs."

In August of 2013, according to the New York Times [3], a leaked draft report from the Intergovernmental IPCC stated in part that:

> "It is extremely likely that human influence on climate caused more than half of the observed increase in global average surface temperature from 1951 to 2010; there is high confidence that this has warmed the ocean, melted snow and ice, raised global mean sea level and changed some climate extremes in the second half of the 20th century."

While the exact wording in this draft may change with further review, observers stated that the substance of these statements would not likely be substantially modified based on reviews.

And here is a summary from the most recent IPCC report, the fifth assessment report (AR5) from the IPCC in 2013 [4]:

> "Human influence has been detected in warming of the atmosphere and the ocean, in changes in the global water cycle, in reductions in snow and ice, in global mean sea level rise, and in changes in some climate extremes. This evidence for human influence has grown since AR4.[2] It is extremely likely that human influence has been the dominant cause of the observed warming since the mid-20th century."

Some impacts from climate change are occurring faster than predicted by computer models, including the summer melting of Arctic ice. The summer Arctic ice sheet reached its smallest extent in 2007 and 2011, opening up two possible Northwest passages from the Atlantic to the Pacific Ocean. Scientists now predict that the Arctic ice could disappear entirely by 2020–2050, far earlier than previously predicted; this would be the first time that the Arctic was free of ice for the last 125,000 years! This would be catastrophic for polar bears, which rely on the ice to search out food. It could also exacerbate climate change on two counts: First, the Arctic water is darker than the ice, so the water would absorb more solar radiation than the ice, increasing the rate of temperature rise. Second, without ice,

[2]AR4 refers to the last IPCC report in 2007.

the Arctic could become the next large source of oil, increasing the likelihood of increased burning of fossil fuels.

This author concludes that human-generated greenhouse gases are most likely causing global warming and that there is a relatively high probability of catastrophic climate change over the next century if we do not act aggressively before 2020 to slow and stop most human-generated greenhouse gas emissions. Given the gravity of the potential impacts of climate change, it seems only prudent to take steps now to avert the possibility of extreme climate change, particularly when such steps will also immediately reduce health threats from local air pollution and will also pave the way for a sustainable energy system that will survive the inevitable decline in fossil fuel production by the end of the twenty-first century after the world passes the "peak oil" production rate and oil extraction costs and gasoline prices rise inexorably.

Since GHG emissions anywhere in the world contribute to the increased level of GHGs accumulating in the atmosphere, any assessment of the impacts of human activity must account for all sources of GHG emissions throughout the entire fuel chain. In the case of gasoline or diesel fuel used in motor vehicles, for example, analysts must calculate the "well-to-wheels" GHG emissions in each step of the process of finding oil, extracting crude oil, transporting that crude oil to refineries, processing the crude oil to make gasoline and diesel fuel, and the transporting those fuels to vehicle fueling stations around the world, in addition to the GHGs actually generated by motor vehicles.

A similar "well-to-wheels" calculation must be made for analyzing the impact of charging batteries on electric vehicles. Since 67 % of all electricity was made in the USA by burning fossil fuels in 2013 (39 % from coal, 27.5 % from natural gas, and 0.67 % from petroleum) according to the Department of Energy's Energy Information Administration (EIA), the analyst must consider the GHG emissions in mining coal, in transporting that coal to the power plants, and finally, the GHGs actually emitted by the power plant. According to the latest EIA projections in their 2015 Annual Energy Outlook [5], this percentage of electricity from fossil fuels is expected to decrease only slightly from 67 % in 2013 to 65 % of all electricity by 2040, meaning that coal (33.7 %), natural gas (31 %), and even oil (0.35 %) will continue to provide most US electricity in the decades ahead,[3] unless the USA seriously attempts to reduce its GHG emissions.

Calculating the GHG emissions for various alternative fuels and vehicles is extremely complex. Fortunately, scientists and engineers at the Department of Energy's Argonne National Laboratory have developed a detailed computer model called GREET [6] that makes all the necessary calculations to estimate local air pollution, GHG emissions, and oil consumption from a wide variety of alternative fuels and alternative vehicles.

[3]According to the 2015 AEO reference case projections, renewable energy will grow only slightly (from 13 % in 2013 to 18 % in 2040, while nuclear power is projected to decrease slightly (from 19.4 to 16.5 %), so the zero-carbon electricity sources will not significantly decrease greenhouse gas emissions from generating electricity in the USA, according to the EIA.

We used the 2012 GREET model to calculate GHG emissions for two separate conditions: near-term emissions through 2035 and a long-term set of multivehicle scenarios over the entire twenty-first century.

8.1 Near-Term (Through 2035) GHG Emissions Without Climate Change Legislation

The estimated average GHG emissions from various alternative vehicles are summarized in the second column of Table 8.1 for the year 2020, based on detailed analysis of regional marginal grid mixes across the USA by the Oak Ridge National Laboratory (See Appendix D for details). GHGs for battery EVs increase in some regions compared to gasoline hybrids, and they decrease in other regions. We have calculated the vehicle-weighted average emissions for BEVs based on the population of vehicles in each electrical region in the USA. On the average, battery EVs in the USA will reduce GHGs by 3.6 % compared to gasoline hybrids such as the Toyota Prius. Fuel cell EVs will decrease GHGs by 24.4 % relative to gasoline hybrids, assuming that all hydrogen is made from natural gas.

All the alternative vehicles listed in Table 8.1 will decrease GHGs compared to a conventional (non-hybrid) gasoline vehicle (last or fourth column), but it is likely that most vehicles will be hybridized by 2020 in order for the auto

Table 8.1 Estimated GHG emissions for various alternative vehicles in 2020, and percentage change in GHGs compared to gasoline hybrids (3rd column) and to conventional gasoline vehicles (last column)

	Vehicle wgt'd average GHGs (g/mile)	% GHGs relative to gasoline HEVs (%)	% GHGs relative to gasoline ICVs (%)
Gasoline ICV	550	62.2	0.0
Gasoline HEV	339	0.0	−38.3
Gasoline PHEV-40	286	−15.7	−48.0
E-85 ICV	398	17.5	−27.5
Diesel ICV	475	39.9	−13.7
Diesel HEV	308	−9.2	−44.0
NGV	488	44.0	−11.2
NG HEV	310	−8.6	−43.7
NG PHEV40	300	−11.5	−45.4
BEV	327	−3.6	−40.6
FCEV	264	−22.2	−52
FC PHEV	286	−15.6	−48

ICV Internal Combustion engine Vehicle; *E-85* mixture of 85 % ethanol in gasoline; *HEV* Hybrid Electric Vehicle; *NGV* Natural Gas Vehicle; *NG* Natural Gas; *BEV* Battery Electric Vehicle; *FCEV* Fuel Cell Electric Vehicle; *PHEV-40* plug-in hybrid electric vehicle with 40 miles all-electric range

companies to meet stringent new corporate average fuel economy (CAFE) standards, so the comparison with gasoline hybrid vehicles (third column) is the most appropriate measure. Thus, car buyers in 2020 will have to choose a conventional hybrid such as the Prius or a fuel cell EV, a battery EV, or plug-in hybrid.

Note that the FCEV reduces GHG emissions by 52 % compared to a conventional gasoline vehicle and by 22.2 % compared to a gasoline hybrid, even though all hydrogen is made from natural gas in the near term. Some advocates of BEVs dismiss FCEVs on the grounds that hydrogen made from natural gas creates some carbon dioxide, the main GHG, which is true. But they fail to note that the GHGs from a FCEV are 22–52 % less than the emissions from any gasoline-powered vehicle.

If no climate change legislation is passed by Congress, then GHG emissions from electrical power plants are unlikely to decrease significantly in the next few decades,[4] meaning that much of the electricity used to charge battery EV and plug-in hybrid batteries will come from burning fossil fuels in the USA. For example, the Department of Energy's Energy Information Administration (EIA) projects very little decrease in fossil fuel combustion in electrical generation plants through 2040 in their 2015 Annual Energy Outlook as described above, with total fossil fuel electricity generation falling only slightly from 67 % in 2013 to 65 % by 2040 which means that GHGs from electricity to charge battery EVs and plug-in hybrids will most likely remain relatively flat over the next 20–25 years without any serious climate change legislation or other incentives to "green the grid."

While Table 8.1 indicates that individual battery EVs will generate slightly more GHG emissions than a fuel cell EVs even if all hydrogen is made from natural gas[5]; that is not the full story. In addition to GHG emissions per vehicle, we need to analyze the *number* of vehicles of each type that are likely to be purchased by future drivers.

As shown in Appendix B, battery EVs will most likely be limited to a maximum of 31 % of all vehicles on the road in the USA with current and planned battery technology, due to the relatively low percentage of small vehicles with short driving ranges that are economically viable for battery power.

[4]The EPA may force some utilities to retire their old coal-burning plants to comply with the Clean Air Act provision on limiting mercury and soot pollution, a process that has already started in 2011 with multiple coal plant closures; however, other plants will continue burning coal or other fossil fuels by adding pollution control devices. Hence it is unlikely that there will be a major shift to cleaner burning power plants without major climate change legislation.

[5]Not all hydrogen is made from natural gas today; one hydrogen fueling station in Fountain Valley, California, already generates the hydrogen from the digester gas at a wastewater treatment plant at the Orange County Sanitation District, which is a true zero-carbon (or even negative-carbon) hydrogen, a waste-to-hydrogen green project. Zero-carbon hydrogen can also be made from landfill gas which does not require the expensive and time-consuming "greening of the grid" that would be necessary before BEVs could reduce GHG emissions.

8.2 Long-Term GHG Emissions with Climate Change Legislation

The previous section assumed that both electricity and hydrogen production continued without any major legislative or regulatory incentives to reduce GHG emissions. For the long term (2020 and beyond), we assume that legislation or other incentives are established to force utilities to replace fossil fuel-based generators with some combination of renewables, nuclear or fossil generators with carbon capture and sequestration or storage[6] (CCS); the assumptions used in the long-term model are described in Appendix A.

Initially, all hydrogen is made from natural gas, either at the fueling station or at a central plant with pipeline or truck delivery; this hydrogen immediately cuts GHGs by 52 % compared to a conventional gasoline car and by 22 % relative to a gasoline hybrid EV. Over time, we assume a gradual transition to lower carbon hydrogen production options such as making hydrogen from ethanol or biomass, wastewater, municipal solid waste, and landfill gas and eventually evolving to the production of hydrogen from coal at a high-efficiency integrated gasification combined cycle plant (IGCC) with carbon capture and storage (CCS), biomass gasification with CCS, and electrolysis of water using zero-carbon renewable or nuclear electricity.

The intermediate option of making hydrogen from biomass or from the anaerobic digester gas at wastewater treatment plants is already operational at the Orange County Sanitation District in Fountain Valley, California. They have installed a stationary molten carbonate fuel cell (MCFC) to produce 250 kW of electricity to help run the wastewater treatment plant. Some of the waste is heated in an anaerobic digester tank that converts the waste into methane and carbon dioxide. The MCFC converts the energy from the methane into electricity at very high efficiency (47 %), which displaces electricity previously purchased from the local utility thereby cutting GHGs, and the waste heat is used to heat the digester tank, reducing natural gas consumption that previously heated these tanks. In addition, some of the methane is converted to hydrogen, which is purified and used as fuel to the hydrogen fueling station in Fountain Valley. This process is already economic (See Chap. 13 and Appendix E), since the wastewater treatment plant saves money on purchased electricity and also reduces costs for natural gas to heat the digester tanks. Since the source of the electricity, heat, and hydrogen is wastewater, this process is totally renewable (as long as people keep flushing their toilets!!). GHGs are cut substantially, since less electricity is generated from fossil fuels, less natural gas burned, and the fuel cell EV running on this waste hydrogen emits no GHGs.[7]

---

[6]Technology is being developed and tested to extract the carbon dioxide, the main greenhouse gas from the power plants, and store or "sequester" the gas underground in salt caverns or depleted natural gas formations.

[7]Some CO_2, the main GHG, is released by the MCFC reformer, but these GHGs are much less than the GHGs displaced by reduced electricity and natural gas consumption, and the GHGs that are saved by replacing conventional vehicles with FCEVs running on this renewable hydrogen with no GHG gas emissions from the vehicle.

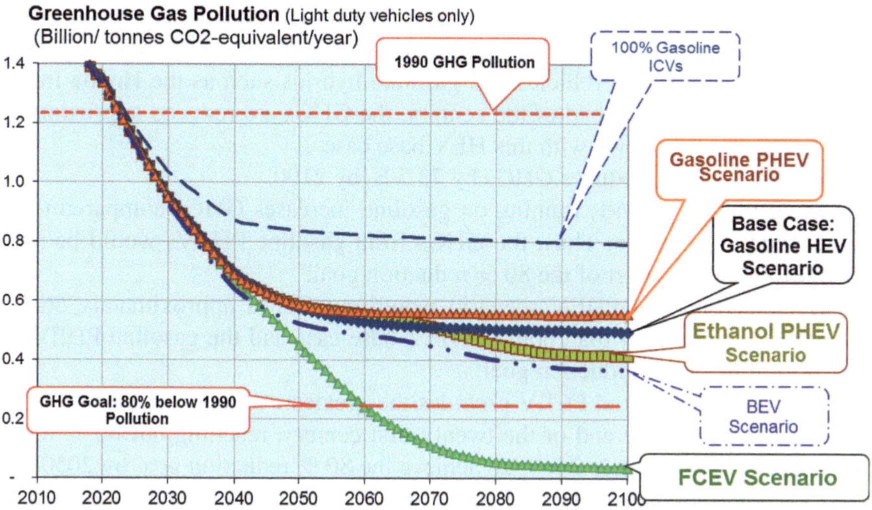

Fig. 8.2 Greenhouse gas emissions from the various alternative vehicle scenarios described in this book compared to the societal goal of cutting GHGs by 80 % below 1990 levels

The greenhouse gases calculated by this model are shown in Fig. 8.2 for the various alternative vehicle scenarios described in Chap. 4. The goal set by the climate change community is to reduce all GHG emissions to at least 80 % below 1990 levels preferably by 2050. The 1990 level of GHG emissions by light-duty vehicles in the USA was approximately 1.23 billion metric tonnes of CO2-equivalent per year, the upper dotted red line in Fig. 8.2. However, even an 80 % reduction may not be sufficient. McKinsey & Company, in their landmark analysis of alternative vehicles in the EU [7], concluded that GHG emissions from the road transportation would have to be reduced by 95 % below 1990 levels in order to achieve an overall societal reduction of 80 %. They estimated that the GHGs from some sectors of the EU economy could not be affordably reduced by 80 %, so the road transportation sector would have to be cut more than 80 %, and they suggested that a 95 % reduction for road transport[8] is feasible, particularly if fuel cell EVs are included in the mix.

The upper dashed line ("100 % Gasoline ICVs") in Fig. 8.2 shows the GHGs without any alternative vehicles.[9] The GHGs are reduced over the century since we assume that even conventional gasoline cars will become more efficient over time with reduced weight, reduced aerodynamic drag, and reduced rolling resistance. Of course, the main avenue for reducing GHGs is via hybridization, which is captured in the base case of primarily gasoline hybrid electric vehicles (HEVs).

[8]Presumably, "road transport" includes both light-duty vehicles (cars, trucks, and SUVs) as well as heavy-duty vehicles including buses and long-haul trucks.

[9]This all-ICV scenario is already obsolete, since HEVs are already capturing a sizeable fraction of global auto sales.

Each alternative vehicle scenario reduces GHGs to some degree.

The "Base case" is essentially a continuation of our current vehicle sales trend of conventional gasoline vehicles and gasoline hybrids such as the Honda Insight and Toyota Prius. By the end of the century, the GHGs are reduced approximately 60 % below the 1990 level with this HEV base case.

The BEV scenario reduces GHGs by 73.5 % by 2100.

Adding plug-in hybrids running on gasoline increases GHGs compared to the base case HEV scenario. Thus, the GHGs from gasoline PHEVs would be 55 % below 1990 level or short of the 80 % reduction goal.

The ethanol/biofuel PHEV scenario would cut GHGs approximately 66.4 % below 1990 level, better than both the HEV base case and the gasoline PHEV, but also short of the 80 % reduction goal.

The hydrogen-powered FCEV provides the greatest reduction of 97.1 % below 1990 GHG level by the end of the twenty-first century, reaching our 80 % reduction goal by approximately 2060. To achieve the 80 % reduction goal by 2050 with FCEVs would require either faster fuel cell EV market penetration and/or a more rapid transition to lower carbon hydrogen production, such as a significant increase in near-term hydrogen production from wastewater treatment plants (See, for example, Chap. 13) or from landfill gas or municipal solid waste, or by electrolysis using nuclear or renewable energy.[10] Other alternatives to augment the FCEV scenario to reach the 80 % reduction goal before 2100 include increased use of mass transit, more ride-sharing or other options to cut down the miles traveled in personal vehicles each year.

Another very promising pathway is to utilize our vast coal resources to make hydrogen by gasifying the coal in an IGCC power plant with CCS in underground cavities. Dr. Sig Gronich, who managed the Department of Energy's hydrogen and fuel cell development programs for many years, points out that this approach might garner strong political support from representatives of the coal industry. Gronich also notes that capturing the carbon in an IGCC plant before combustion of the coal is less costly than waiting until after combustion to try to capture the CO_2 in the smokestack of a conventional coal-burning power plant,[11] making CCS more feasible [8]. While IGCC generally refers to using the synthetic gas or syngas [a mixture of carbon monoxide (CO) and hydrogen] from the coal gasifier to run a gas turbine to generate electricity, the CO in the syngas can be converted to additional hydrogen by a water gas shift reactor ($CO + H_2O = H_2 + CO_2$), providing a good source of "green" hydrogen for fueling FCEVs as well as to make "green" electricity in a stationary fuel cell. This coal IGCC approach to producing hydrogen has these major attributes:

[10]However, electrolytic hydrogen requires significant water which is also in short supply in many parts of the world, so making hydrogen by splitting water may be limited in the future, unless technology can be developed to electrolyze seawater.

[11]CO_2 concentrations might be 15–50 % precombustion, but only 5–15 % in the smokestack after combustion in a conventional coal boiler making CO_2 separation much easier precombustion in an IGCC plant.

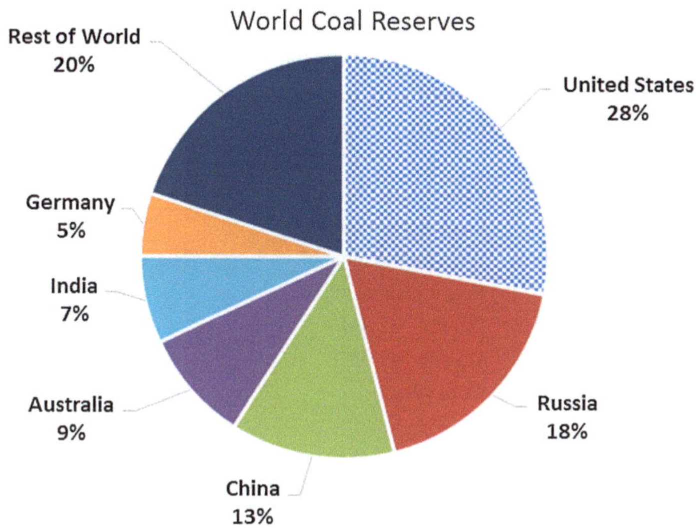

Fig. 8.3 Estimated world coal reserves by country

- Utilizes a major source of indigenous fuel (see Fig. 8.3)
- Has strong political support to find an environmentally sound use of coal.
- Permits production of both green electricity and green hydrogen for transportation.
- Hydrogen provides a good method for storing energy to ameliorate the impact of grid outages and intermittent power from renewables.

In short, hydrogen can provide an excellent pathway for providing an ultra-clean transportation fuel and an approach to providing clean affordable electricity.

Relying on hydrogen made from coal would flip the tables on the Middle East with respect to energy reserves. The USA has approximately 28 % of the world's known coal reserves, while the entire Middle East and North Africa are estimated to have only 0.108 % of total coal reserves [9].

In terms of GHG emissions, CCS in underground salt caverns does not sequester all the carbon dioxide separated from the coal gas stream; Argonne National Laboratory assumes 80 % capture in their GREET model. The resulting GHG emissions from a FCEV from various energy sources from the GREET model is summarized in Table 8.2. A FCEV running on hydrogen from coal-based IGCC with CCS would generate approximately 61.6 g of CO_2-equivalent per mile, which is 57.5 % greater than the GHGs from a FCEV running on electrolytic hydrogen from renewable or nuclear electricity (the lowest GHG option), but less than half the GHG for a FCEV running on hydrogen made from natural gas, the current lowest cost option.

Table 8.2 GHG emissions in grams of CO_2-equivalent per mile for a FCEV running on different sources of hydrogen and percentage increase over electrolytic hydrogen

	g/mile	%
Nuclear and renewable electrolysis	39.1	0.0
Farmed trees ethanol	49.1	25.6
Biomass gasification	53.5	36.8
Coal IGCC with CCS	61.6	57.5
Miscanthus ethanol	71.3	82.4
Corn stover ethanol	78.4	100.5
Switchgrass ethanol	82.5	111.0
Forest residue ethanol	94.8	142.5
Central NG SMR	143.8	267.8
On-site NG SMR	151.4	287.2
Corn ethanol	170.8	336.8

References

1. "Global Greenhouse Gas Reference Network," Earth Systems Research Laboratory, Global Monitoring Division, National Oceanic and Atmospheric Administration, Available at: http://www.esrl.noaa.gov/gmd/ccgg/trends/.
2. Climate Change Science: An Analysis of some key Questions, by the Committee on the Science of Climate Change, National Research Council of the National Academy of Sciences, ISBN 0-309-07574-4 (2001), available at http://www.nap.edu/catalog.php?record_id=10139.
3. Justin Gillis, "Climate Panel Cites Near Certainty on Warming," The New York Times, August 19, 2013, available at: http://www.nytimes.com/2013/08/20/science/earth/extremely-likely-that-human-activity-is-driving-climate-change-panel-finds.html?adxnnl=1&src=me&adxnnlx=1377029122-olLP3nIPAdlkPrb/xm7M2Q&_r=0.
4. IPCC, 2013: Summary for Policymakers. In: Climate Change 2013: The Physical Science Basis. Contribution of Working Group I to the Fifth Assessment Report of the Intergovernmental Panel on Climate Change [Stocker, T.F., D. Qin, G.-K. Plattner, M. Tignor, S.K. Allen, J. Boschung, A. Nauels, Y. Xia, V. Bex and P.M. Midgley (eds.)]. Cambridge University Press, Cambridge, United Kingdom and New York, NY, USA. Available at: http://www.climatechange2013.org/images/report/WG1AR5_SPM_FINAL.pdf.
5. The Annual Energy Outlook 2015, by the DOE's Energy Information Administration, available at: http://www.eia.gov/oiaf/aeo/tablebrowser/.
6. M. Q. Wang, "The Greenhouse gas, regulated emissions and energy use in transportation (GREET)," Energy Systems Division, Argonne National Laboratory, available at: http://greet.es.anl.gov/.
7. McKinsey & Company, "A Portfolio of power-trains for Europe: a fact-based analysis: The Role of Battery Electric Vehicles, Plug-in Hybrids and Fuel Cell Electric Vehicles" 2007, available at: http://cleancaroptions.com/html/a_portfolio_of_power_trains_for_europe_a_fact_based__analysis.pdf.
8. Private communication, Dr. Sig Gronich, January 23, 2015.
9. U.S. Energy Information Agency, available at http://www.eia.gov/todayinenergy/detail.cfm?id=2930.

Chapter 9
Projected Oil Consumption for Alternative Vehicles

Abstract This chapter summarizes the alternative vehicle petroleum consumption findings from this book. We analyze the petroleum consumption for all US alternative vehicles and fuels for two time periods similar to the greenhouse gas periods covered in Chap. 8: (1) A near-term "Business-as-usual" condition through 2020, and (2) A longer-term assessment over the twenty-first century. Our goal is to reduce oil consumption to a level that all oil imports can be supplied by non-OPEC nations and preferably by friendly nations from the Americas.

9.1 Near-Term Oil Consumption Projections

In the near-term (through 2035), reducing oil consumption from the transportation sector will be difficult without a mandate such as reducing GHG emissions to drive a significant shift to alternative vehicles and/or fuels. Table 9.1 shows the calculated petroleum consumption (in BTUs/mile)[1] of the various alternative vehicles from the Argonne GREET model. To cut oil consumption by 50 % relative to a gasoline HEV will require one of the following alternative vehicle choices (or combinations thereof):

- E-85 HEVs or E-85 ICVs
- NGVs
- NG HEVs
- BEVs
- FCEVs, or
- NG or FC PHEVs.

[1]These petroleum consumption values are based on the LDV-weighted marginal electricity grid mixes that will affect the PHEV oil consumption; petroleum consumption will be slightly less using the *average* grid mixes for BEVs and PHEVs (see Appendix D for details on marginal vs. average electricity mixes).

C.E. (Sandy) Thomas, *Sustainable Transportation Options for the 21st Century and Beyond*, DOI 10.1007/978-3-319-16832-6_9

Table 9.1 Petroleum consumption and percentage changes compared to a gasoline hybrid for various alternative vehicles circa 2020

	Average petroleum consumption (BTU/mile)	% change relative to gasoline HEV
Gasoline ICV	5,961	62.8
Gasoline HEV	3,661	0.0
E-85 ICV	2,214	−39.5
E-85 HEV	1,692	−53.8
Gasoline PHEV-40	1,727	−52.8
Diesel ICV	5,146	40.5
Diesel HEV	3,326	−9.2
NGV	581	−84.1
NG HEV	82	−97.8
NG PHEV-40	326	−91.1
BEV	329	−91.0
FCEV	697	−81.0
FC PHEV	484	−86.8

These are oil consumption reductions *per vehicle*. Without a major project such as a climate change initiative, it is unlikely that enough of these vehicles would be purchased to make even a minor dent in our voracious appetite for petroleum. As described in Chap. 2, US oil imports peaked at 2.5 billion barrels in 2005, before falling back to 2.48 billion barrels in 2008 due to the great recession, but 2013 imports were back to 2.24 billion bbl/year.

From Chap. 2, our goals for reducing oil consumption are cutting the LDV oil use to 1.29 billion bbl/year so that the USA could satisfy all oil needs from non-OPEC nations and preferably to 1.1 billion bbl/year such that we could supply all petroleum requirements from friendly nations on the American continent in a crisis. These two goals require reductions between 950 million bbls/year (non-OPEC goal) and 1.14 billion bbl/year (Americas-only goal).

Table 9.2 summarizes the results of this analysis. The first column shows the estimated oil consumption (BTUs/mile) from the Argonne National Laboratory GREET model for each alternative fuel vehicle (AFV).

The last four columns show the number of AFVs required (in millions) and the number of years to achieve those sales *assuming 100 % sales of the AFVs every year* (!). Thus, 70 million BEVs would be required to meet the non-OPEC goal and 86 million BEV sales to meet the American-only goal with BEVs. If 100 % of all LDV sales (7.2 million/year assumed here) were BEVs, it would take between 10 and 12 years of 100 % BEV sales to achieve these goals.

This clearly illustrates that it will take many years if not decades to achieve these petroleum reduction goals, as shown in the next section with a more rational deployment of AFVs over time.

Table 9.2 Oil consumption for alternative-fueled vehicles (AFVs) from the Argonne GREET model, annual oil savings relative to conventional (non-hybrid) gasoline vehicles and millions of AFVs required to meet oil reduction goals, and resulting years of **sales of 100 % AFVs** required to meet oil reduction goals

	Average petroleum consumption (BTU/mile)	Annual consumption for one million AFVs (Mbbl/yr)	Annual oil savings from one million AFVs (Mbbl/yr)	Millions of AFVs required to meet oil reduction goals		Years of 100 % AFV sales to meet oil reduction goals	
				Non-OPEC	Americas-only	Non-OPEC	Americas-only
Gasoline ICV	5,961	14.54					
Gasoline HEV	3,661	8.93	5.61	418	450	58	62
E-85 ICV	2,214	5.40	9.14	256	276	36	38
E-85 HEV	1,692	4.13	10.42	225	242	31	34
Gasoline PHEV-40	1,727	4.21	10.33	227	244	32	34
Diesel ICV	5,146	12.55	1.99	1,718	1,269	164	176
Diesel HEV	3,326	8.11	6.43	365	393	51	55
NGV	581	1.42	13.13	179	192	25	27
NG HEV	82	0.20	14.34	163	176	23	24
NG PHEV-40	326	0.80	13.75	171	184	24	26
BEV	329	0.80	13.74	171	184	24	26
FCEV	697	1.70	12.84	183	197	25	27
FC PHEV	484	1.18	13.36	175	189	24	26

9.2 Long-Term Oil Consumption Projections

The projected oil consumption over the twenty-first century is shown in Fig. 9.1 for the various alternative vehicle scenarios. This is the same model with the same input assumptions used in Chap. 8 for the greenhouse gas projections. Most of the alternative vehicle scenarios reach our twin goals of utilizing only non-OPEC oil by the 2040–2050 time period.

Gasoline PHEVs are superior to HEVs regarding oil consumption, since they derive part of their power from grid electricity, which does not use significant petroleum with the exception of Alaska, Hawaii, and parts of the northeast USA.

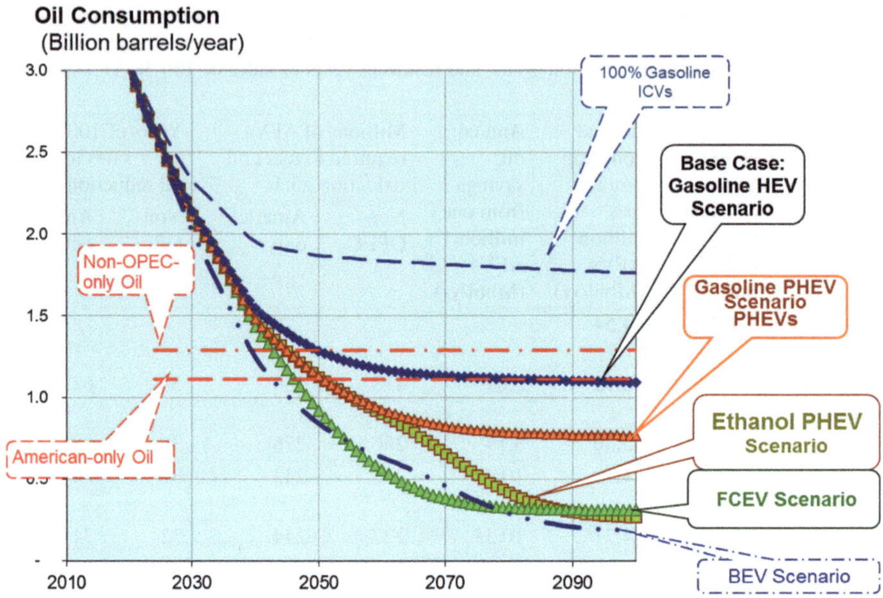

Fig. 9.1 Projections of US petroleum consumption over the twenty-first century with the same input assumptions used in Chap. 8 for estimating greenhouse gas emissions over the century

Table 9.3 Year to reach petroleum reduction goals

	Year to reach goal	
	Non-OPEC	Americas-only
FCEV	2043	2047
BEV	2040	2043
Base case HEV	2050	2080
EtOH PHEV	2045	2051
Gasoline PHEV	2045	2051

We conclude that BEVs are the best option to reduce petroleum consumption in the twenty-first century, although FCEVs are only 3 to 4 years behind the BEV oil reductions.[2] As summarized in Table 9.3, the FCEV scenario reaches the non-OPEC goal by 2043, while the BEV scenario reaches that goal by 2040; the ethanol and gasoline PHEV scenarios by 2045, and the base case by 2050.

Similarly, the FCEV scenario reaches the Americas-only oil goal by 2047, while the BEV scenario achieves that goal four years earlier; the ethanol and

[2]As shown in Chap. 11, natural gas hybrids (NG HEVs) reduce oil consumption more than FCEVs with our assumption of relatively rapid deployment of NG HEVs, but NG HEVs would increase urban air pollution health costs and could not meet our GHG reduction targets.

gasoline PHEV scenarios 4 years later, and the based case gasoline HEV scenario reaches the Americas-only 33 years after the FCEV scenario.

The BEV scenario does reduce petroleum consumption slightly more than the FCEV scenario by the end of the century, primarily because the BEV scenario is predominantly ethanol PEHVs by the end of the century, since BEVs can only capture a maximum of 31 % of the LDV market according to our analysis, and we used ethanol PHEVs as the alternative to the BEV in the BEV scenario to give the best possible outcome to the BEV scenario (if the other non-BEVs in the BEV scenario were gasoline PHEVs instead of ethanol PHEVs, then the BEV scenario would have had much greater petroleum consumption).

Chapter 10
Local Air Pollution

Abstract This chapter summarizes the estimated local (urban) air pollution emissions for each of the alternative vehicles analyzed in this book. At the end of the twenty-first century in the USA, local air emissions were the main motivation for developing cleaner, alternative vehicles to reduce smog in major cities, driven primarily by the smog in Los Angeles. Urban air pollution is still a major issue in many urban cities around the world, although climate change and reliance on imported oil from unstable regions of the world are now the main drivers for developed nations to develop and deploy alternative fuels and vehicles.

The urban[1] air pollution for the alternative vehicles is compared with conventional diesel and gasoline vehicles in mg/mile in Table 10.1 as calculated by the Argonne National Laboratory GREET 2014 model [1]. From Chap. 11, we know that neither natural gas- nor diesel-fueled vehicles will be sustainable over the long run due primarily to their excessive GHG emissions. But, according to the GREET-2014 model, natural gas and diesel vehicles, particularly in the hybrid mode, would reduce local urban air pollution compared to conventional gasoline vehicles.

To quantify the cost of urban air pollution, we can assign an annual societal cost for each pollutant as shown in Table 10.2, which summarizes the average societal costs (mostly health cost of the pollutants) from papers by Delucchi [2] and by Thomas [3].

We use the average of these two estimates to gauge the total air pollution costs for each alternative vehicle as shown in Fig. 10.1, which shows that diesel ICVs and NGVs will both reduce the societal costs of air pollution compared to gasoline ICVs, and diesel and natural gas HEVs will both reduce societal costs compared to gasoline HEVs, but both diesel and natural gas HEVs will have larger urban air

[1]The urban air emissions exclude upstream emissions in fuel production that are outside the local urban airshed and would therefore not contribute to the formation of urban smog.

C.E. (Sandy) Thomas, *Sustainable Transportation Options for the 21st Century and Beyond*, DOI 10.1007/978-3-319-16832-6_10

Table 10.1 Urban air pollution in mg/mile for various alternative vehicles compared to conventional gasoline vehicles

	Urban air pollution (mg/mile)				
	NOx	PM-10	PM-2.5	VOC	SOx
Gasoline ICV	154.1	17.8	12.2	159.2	32.7
Diesel ICV	125.1	11.2	8.8	65.7	24.4
Gasoline HEV	122.0	14.3	10.2	159.2	32.7
Diesel HEV	105.2	9.9	8.0	51.0	18.3
NGV	117.0	6.3	5.7	77.4	10.6
NG HEV	94.6	6.0	5.5	46.7	7.2
FCEV	16.5	6.7	6.6	1.9	4.6
BEV	56.9	10.1	7.2	1.4	111.5

Table 10.2 Possible societal costs of urban air pollution in $/tonne of pollutant

	$/tonne		
	Delucchi	Thomas	average
VOC: urban	–	7510	7510
CO: urban	55	1677	866
NOx: urban	11,885	13,297	12,591
PM10: urban	13,385	36,835	25,110
PM2.5: urban	120,085	118,552	119,319
SOx: urban	50,280	21,873	36,076

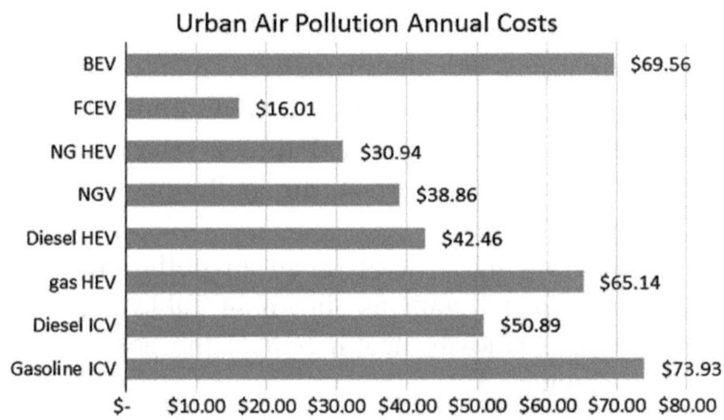

Fig. 10.1 Total per vehicle societal costs estimated for urban air pollution from alternative vehicles

societal costs than hydrogen-powered FCEVs.[2] The FCEV generates the least costly urban air pollution when all hydrogen is made from natural gas.

Diesel and natural gas HEVs would have lower urban air costs than BEVs. The high BEV air pollution costs are due primarily to the SOx emissions from local power plants in the urban airshed used to charge EV batteries.[3]

---

[2]Diesel HEVs urban air costs would be 2.7 times larger than FCEV costs, and natural gas HEV air costs are 4.1 times larger than FCEV urban air costs.

[3]SOx costs account for 69 % of the total BEV urban air societal costs.

References

1. M. Wang, "The Greenhouse gas, Regulated Emissions, and Energy Use in Transportation (GREET) Model", available at: https://greet.es.anl.gov/.
2. M. Delucchi, "Environmental Externalities of Motor-Vehicle Use in the U.S.", Journal of Transport Economic and Policy, ISSN 0022 5258, Vol. **34**, Part 2, May 2000, pp. 135-168.
3. C. Thomas, "Transportation Options in a Carbon-constrained world: Hybrids, Plug-in hybrids, Biofuels, Fuel Cell Electric Vehicles, and Battery Electric Vehicles," International Journal of Hydrogen Energy **34** (2009) pp 9279-9296.

Chapter 11
Natural Gas and Diesel Hybrid Electric Vehicles

Abstract The main thesis of this book is that we must replace the venerable internal combustion engine with an electric motor if we are to build a truly sustainable transportation system. But, some analysts suggest that natural gas or diesel vehicles powering internal combustion engines could make a contribution to reduce GHG emissions, oil imports, and local air pollution. This chapter compares these two fossil fuels with the other EV options discussed earlier. In order to make the strongest case for diesel and natural gas fuels, we assume that both will be used in hybrid electric vehicles to maximize their efficiency and minimize their greenhouse gas emissions and petroleum consumption, even though no automobile company is known to be developing hybrid vehicles running on diesel fuel or natural gas.

11.1 Greenhouse Gas Emissions

Table 11.1 compares the GHG emissions and oil consumption per vehicle for the various alternative vehicles considered in this book; these GHG calculations are based on the marginal grid mixes for the USA as calculated by the Oak Ridge National Laboratory for the 13 electricity regions in the USA[1] plus a "14th region" consisting of Alaska and Hawaii. Our baseline comparison for GHG emissions is the gasoline HEV, which is the best commercial vehicle available today at 286 g/mile of CO_2-equivalent GHG emissions. Both the diesel HEV and the NG HEV have lower GHG emissions compared to the gasoline HEV. The bold emphasized areas in Table 11.1 indicate alternative vehicles with GHG emissions and/or petroleum consumption greater than those from the gasoline HEV and therefore not suitable to meet our societal goals. We therefore ran only two more alternative vehicle scenarios (in addition to those scenarios considered in Chaps. 8 and 9) to test these two options that would reduce GHGs relative to a gasoline HEV: the diesel HEV and the NG HEV scenarios.

[1]See Fig. D.2 in Appendix D.

C.E. (Sandy) Thomas, *Sustainable Transportation Options for the 21st Century and Beyond*, DOI 10.1007/978-3-319-16832-6_11

Table 11.1 Greenhouse gas emissions and oil consumption of alternative vehicles per vehicle in the 2020 time period and percentage of changes compared to the gasoline HEV

	Average petroleum consumption (BTU/mile)	% change relative to gasoline HEV (%)	Average greenhouse gas emissions (g/mile)	% change relative to gasoline HEV (%)
Gasoline ICV	5,961	62.8	550	62.2
Gasoline HEV	3,661	0.0	339	0.0
E-85 ICV	2,214	−39.5	398	17.5
E-85 HEV	1,692	−53.8	305	−10.0
Gasoline PHEV-40	1,727	−52.8	286	−15.7
Diesel ICV	5,146	40.5	475	39.9
Diesel HEV	3,326	−9.2	308	−9.2
NGV	581	−84.1	488	44.0
NG HEV	82	−97.8	310	−8.6
NG PHEV-40	326	−91.1	300	−11.5
BEV	329	−91.0	327	−3.6
FCEV	697	−81.0	264	−22.2
FC PHEV	484	−86.8	286	15.6

In order to run a simulation, we have to choose the rate of introduction of the alternative vehicle into the marketplace. In the case of diesel HEVs, there is very little technical innovation required, although no automobile company has yet introduced a commercial diesel HEV. In terms of fuel infrastructure, diesel fuel is already available in most parts of the USA and even more ubiquitous in Europe; diesel HEVs might be introduced faster than hydrogen-powered FCEVs.

Natural gas HEVs are in a similar position, with no new vehicle technology required. Natural gas infrastructure is less widespread than diesel infrastructure in the USA. Furthermore, adding more natural gas infrastructure could be almost as costly as adding hydrogen infrastructure, since a natural gas fueling station has many of the same expensive components as a hydrogen fueling station, including compressors, high-pressure storage tanks, and dispensers.

We postulate faster introduction of both diesel and natural gas HEVs than FCEVs as shown in Fig. 11.1 to maximize the potential of diesel and natural gas hybrids. Both sales are represented by logistic functions with the diesel HEV and NG HEV sales starting later[2] but reaching 50 % of sales 13 years earlier[3] than the FCEV sales, both eventually reaching 99 % of all LDV sales by the end of the century.

[2]Since no automobile companies have begun developing diesel or natural gas hybrids, while they are developing and planning to sell or lease commercial FCEVs in the 2014–2017 time period.

[3]We assume that diesel and natural gas hybrids could ramp up somewhat faster than FCEVs since there is already extensive diesel fueling infrastructure and some natural gas infrastructure in place in the USA, while hydrogen infrastructure is minimal to date.

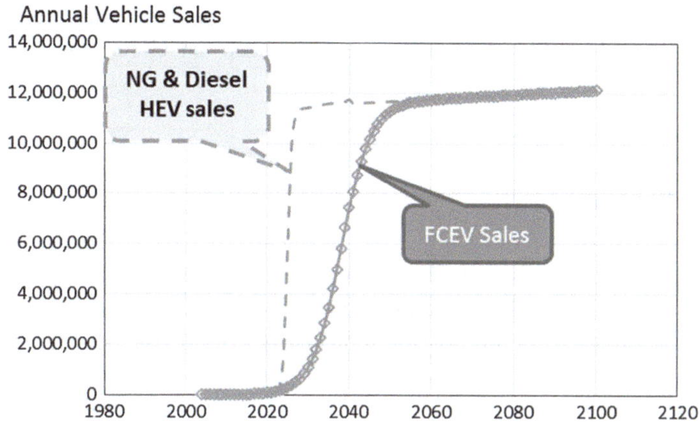

Fig. 11.1 Postulated sales logistics functions for diesel and NG HEVs compared to FCEV sales

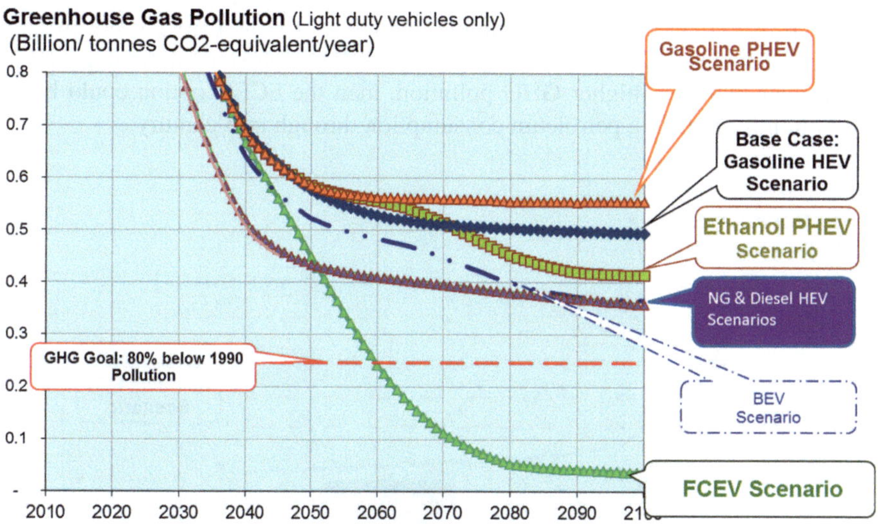

Fig. 11.2 Greenhouse gas emissions including natural gas HEV and diesel HEV scenarios

The resulting GHG emissions over the twenty-first century are shown in Fig. 11.2 for these scenarios. The natural gas HEV scenario has slightly lower GHGs than the diesel HEV scenario over much of the century. The GHGs for these two HEV scenarios are less than the base case with gasoline HEVs, and the curves are nearly identical in Fig. 11.2 (the natural gas HEV scenario GHG level is slightly less than the diesel HEV scenario, but the lines overlap on the figure). Figure 11.2 shows that the hydrogen-powered FCEV is the only option that can reach our target of reducing GHGs to 80 % below 1990 levels, and neither natural gas nor diesel fuel could permit us to even approach our GHG goal, even if these fossil fuels were used exclusively in HEVs.

11.2 Oil Imports

The model projections for petroleum consumption over the twenty-first century are shown in Fig. 11.3. If petroleum reduction was our only goal, then the natural gas HEV scenario is the best option, nearly eliminating all oil consumption by mid-century; however, the natural gas HEV could not sufficiently reduce GHG emissions as shown in Fig. 11.2. The FCEV scenario reaches our goal of obtaining all oil from non-OPEC nations in a future crisis by 2029, with the BEV scenario achieving that goal one year earlier. The FCEV scenario reaches the goal of oil imports from only the American continent by 2047 and natural gas HEV scenario by 2033 as summarized in Table 11.2. However, this result is due primarily to our assumption that natural gas HEVs would ramp up much faster than FCEVs as shown in Fig. 11.1. If the automobile companies follow their current plans to deploy FCEVs in the 2015–2020 time period while not pursuing natural gas HEVs due to their relatively higher GHG pollution, then the FCEV option could be the best option for reducing petroleum consumption through mid-century.

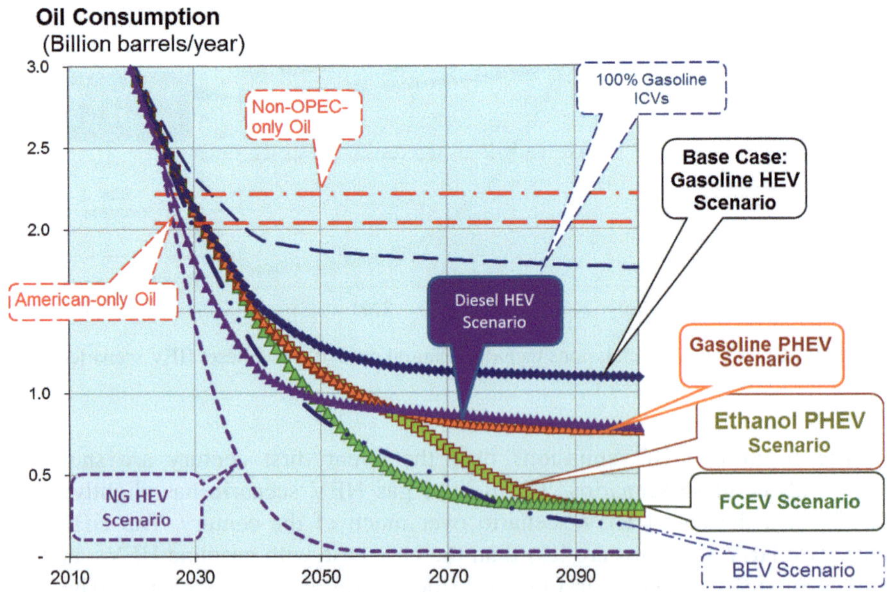

Fig. 11.3 Petroleum consumption for the various scenarios over the twenty-first century including the natural gas HEV and diesel HEV scenarios

Table 11.2 Years to reach our petroleum consumption reduction goals

	Year to reach goal	
	Non-OPEC	Americas only
FCEV	2043	2047
BEV	2040	2043
Base case HEV	2050	2080
EtOH PHEV	2045	2051
Gasoline PHEV	2045	2051
Diesel HEV	2038	2041
NG HEV	2031	2033

Table 11.3 Local air pollution annual costs and relative cost ratios compared to costs for a FCEV

	Societal pollution cost ($/year)	Cost relative to FCEV
Gasoline HEV	65.14	4.07
Diesel HEV	42.46	2.65
NG HEV	30.94	1.93
FCEV	16.01	1.00

The diesel HEV scenario is better than the base case gasoline HEV scenario in terms of reduced petroleum consumption and almost as good as the gasoline PHEV scenario as shown in Fig. 11.3. The natural gas HEV scenario nearly eliminates all petroleum consumption by the end of the century. In terms of meeting our two goals of reducing oil consumption so that in an emergency we could supply all of our needs from non-OPEC nations, the natural gas HEV and FCEV scenarios achieve this goal by 2031 and 2043, and the FCEV scenario reaches the goal of supplying all oil from friendly American nations by 2047 while the natural gas HEV scenario achieves that goal by 2033, or 14 years earlier as summarized in Table 13.2. The diesel HEV scenario reaches the non-OPEC goal by 2038 and the Americas-only goal by 2041.

11.3 Local Air Pollution

We estimated the societal costs of local air pollution in Chap. 10. The estimated annual urban air pollution costs for diesel and natural gas HEVs are compared with the costs for FCEVs in Table 11.3 taken from Table 10.2. We estimate that a natural gas HEV would increase local air pollution costs by a factor of almost two compared to a FCEV, while a diesel HEV would increase local air pollution costs by a factor of 2.65 compared to a FCEV.

11.4 Conclusions

11.4.1 Diesel HEV Conclusions

We conclude that a diesel HEV scenario would

- Increase local air pollution costs by a factor of 2.65 compared to a FCEV
- Reduce GHG emissions more than a gasoline HEV; the Diesel HEV scenario would cut GHGs by 70.1 % below 1990 levels by the end of the century, more than the HEV base case scenario at a 60 % reduction, but well short of our goal of cutting GHGs by 80 % below 1990 levels and much less than the FCEV scenario that will cut GHG emissions by 97.1 % below 1990 levels by the end of the century.

11.4.2 Natural Gas HEV Conclusions

A natural gas HEV scenario would

- Increase local air pollution costs by a factor of 1.9 compared to a FCEV scenario
- Reduce petroleum consumption more than any other scenario, allowing petroleum imports from non-OPEC nations by 2031.
- Reduce GHG emissions more than a gasoline HEV; the natural gas HEV scenario would cut GHGs by 71.7 % below 1990 levels by the end of the century, well short of our goal of cutting GHGs by 80 % below 1990 levels and much less than the FCEV scenario that will cut GHG emissions by 97.1 % below 1990 levels.

Chapter 12
State and International Alternative Vehicle Activities

Abstract We conclude from our detailed analysis of alternative vehicles described in previous chapters that hydrogen-powered fuel cell electric vehicles (FCEVs) offer the best potential for a truly sustainable transportation system. In addition to Federal funding for hydrogen and FCEV developments, several US states have developed their own alternative vehicle programs. Several other nations have made significant commitments to develop robust FCEV programs, which should give the automobile companies sufficient hydrogen fueling opportunities to begin large-scale production and deployment of FCEVs.

12.1 Federal Fuel Cell R&D

The first Secretary of Energy in the Obama administration, Dr. Steven Chu, attempted to eliminate all funding for the Federal hydrogen and fuel cell R&D program in his first budget submission to Congress. Fortunately, at that time Senator Byron Dorgan of North Dakota was the Chairman of the Senate Energy and Water Appropriations Subcommittee that funds fuel cell R&D in the Department of Energy and a strong supporter of fuel cell electric vehicle (FCEV) programs, and the senator was able to maintain most of the funding for the program, and senior staff responsible for the hydrogen and FCEV programs at the Department of Energy were able to maintain these programs notwithstanding the lack of support from their boss.

C.E. (Sandy) Thomas, *Sustainable Transportation Options for the 21st Century and Beyond*, DOI 10.1007/978-3-319-16832-6_12

12.2 State Fuel Cell Initiatives

Several states have active fuel cell programs that have helped to bridge the occasional wavering Federal support for FCEVs and hydrogen infrastructure in the USA. For a detailed report on US state fuel cell and hydrogen activities, see Ref. [1].

12.2.1 California

The state of California has maintained active and vigorous bipartisan[1] support for developing hydrogen infrastructure and FCEVs as one main element of their zero-emission vehicle (ZEV) program started in 1990 that requires all auto companies to offer ZEVs for sale in the state, purchase credits from other auto companies selling ZEVs, or pay a fine. Both BEVs and FCEVs qualify as ZEVs, but FCEVs get more credits because of their longer range.

California is also home to the California Fuel Cell Partnership (CFCP), which is an industry/university/government group that was formed in 1999 by the California Air Resources Board, the California Energy Commission, Ballard fuel cell company, two auto companies (Ford and DaimlerChrysler), and three oil companies[2]; CFCP now has 38 members comprised of eight auto companies,[3] several hydrogen fuel suppliers, universities, two national laboratories, two transit agencies, and state and Federal government organizations such as DOE and EPA.

Other states that have active hydrogen and fuel cell programs as described in the US state fuel cell and hydrogen report [1] include:

- Connecticut
- New York
- Ohio
- South Carolina
- Hawaii
- Massachusetts
- New Jersey
- Pennsylvania
- Rhode Island

[1]Both Republican Governor Arnold Schwarzenegger and his Democratic successor Governor Jerry Brown have been strong supporters of hydrogen and fuel cell electric vehicles.

[2]BP, ChevronTexaco & Shell Hydrogen.

[3]Chrysler, Daimler, GM, Honda, Hyundai, Nissan, Toyota, and VW.

12.3 International Fuel Cell Programs

One leading indicator of a nation's commitment to hydrogen and FCEVs is the number of hydrogen fueling stations in operation and planned.

12.3.1 Japan

Japan has an aggressive program to introduce and support hydrogen-powered FCEVs. Japan had 17 hydrogen fueling stations in 2013, with commitments from the government, three car companies (Toyota, Honda, and Nissan), and ten fuel suppliers to install 100 hydrogen stations by 2016, when the car companies are scheduled to introduce commercial FCEVs [2]. Japan has approximately 74 million registered vehicles, compared to 250 million for the USA, so 100 hydrogen stations in Japan would be equivalent to 340 hydrogen stations in the USA on a per vehicle basis.

12.3.2 Germany

Germany also has ambitious plans to deploy 100 hydrogen stations by 2017, up from the 30 stations in 2013 (of which only 7 were open to the public), with a goal of 400 hydrogen stations by 2023 [3]. Daimler and Linde (a major hydrogen gas supplier) have teamed up to build 20 new hydrogen stations in the 2015 time period.

12.3.3 Scandinavia

Five Scandinavian countries (Sweden, Norway, Finland, Iceland, and Demark) formed the Scandinavian Hydrogen Highway Partnership (SHHP) in 2006 [4]. Their plan is to build 15 hydrogen production "nodes" with 30 satellite hydrogen fueling stations. Seven hydrogen stations were operational in 2014.

12.3.4 UK

The UK government is actively supporting funding for the building of hydrogen fueling stations. As of 2014, there were 6–8 hydrogen fueling stations in the UK, with funding available to build 4–7 new stations [5].

12.3.5 South Korea

South Korea had 13 hydrogen fueling stations in 2012, with plans to have 43 by 2015, 168 by 2020, and 500 by 2030 [6].

References

1. S. Curtin & J. Gangi, "State of the States: Fuel Cells in 2014," 5th Edition, Fuel Cell Technologies Office, DOE, written by Breakthrough Technologies Institute, Washington D.C. November 2014.
2. Y. Sone, "Fuel cell vehicle demonstration and hydrogen infrastructure project in Japan," The Research Association of Hydrogen Supply/ Utilization Technology (HySUT), June 16, 2014. Available at: http://www.iea.org/media/workshops/2014/asiahydrogenworkshop/SessionIISone HysutFCEVandH2infrastructuredemonstrationinJapan.pdf.
3. NOW (National Organization for hydrogen and fuel cell technology) 2013 Annual Report, available at: http://www.now-gmbh.de/fileadmin/user_upload/RE_Publikationen_NEU_2013/ Publikationen_NOW_Berichte/NOW_Annual_Report__2013.pdf.
4. S. Wolf et al., "Advancing the Nordic Hydrogen Innovation Network," May 2010.
5. The UK Hydrogen and Fuel Cell Association (UKHFCA) available at: http://www.ukhfca.co.uk/.
6. J.kim, "Recent Achievements in hydrogen and fuel cells in Korea," International hydrogen energy development, 2013.available at: http://hydrogenius.kyushu-u.ac.jp/cie/event/ihdf2013/pdf/2-3kim.pdf.

Chapter 13
Waste to Hydrogen

Abstract What if hydrogen could be made from waste products, without using any fossil fuels or creating any local air pollution, while simultaneously *reducing* net greenhouse gas emissions and eliminating all petroleum consumption from the transportation sector? Such a project is feasible today using the waste gases generated at landfills or waste water treatment plants. This chapter describes and analyzes the economics of such a waste-to-hydrogen project that is not only feasible today, but makes a 10-year internal rate of return on the project investment of 17.9 % in California, and up to 18.3 % in New York (Long Island) even though New York has not made the commitment to build a significant hydrogen refueling system similar to the California efforts, making this type of hydrogen station a profitable investment in the near term; the economic returns are attractive primarily by displacing electricity (and to a lesser degree heat) to run the wastewater treatment plant; hydrogen sales for fueling vehicles is a minor fraction of the positive cash flows for this system.

13.1 Tri-Gen System Description

Figure 13.1 shows the block diagram of the Tri-Gen system installed at a municipal wastewater treatment plant (WWTP)[1] that produces electricity, heat, and hydrogen.[2]

The waste products from the sanitary sewer are first concentrated in settling tanks, and the resulting sludge is piped to a set of anaerobic digester tanks. Microbes in these heated tanks then convert the solid waste into a gas stream that

[1]Such a Tri-Gen system is operating today at the Orange County Sanitation District in Fountain Valley, California, where hydrogen is currently available to refuel FCEVs; however, the Fountain Valley system uses a more expensive high-temperature molten carbonate fuel cell (MCFC) system instead of the PEM fuel cell system proposed here.

[2]The heat exchangers to extract heat from the steam methane reformer (SMR) system and from the fuel cell system are not shown in Fig. 13.1.

C.E. (Sandy) Thomas, *Sustainable Transportation Options for the 21st Century and Beyond*, DOI 10.1007/978-3-319-16832-6_13

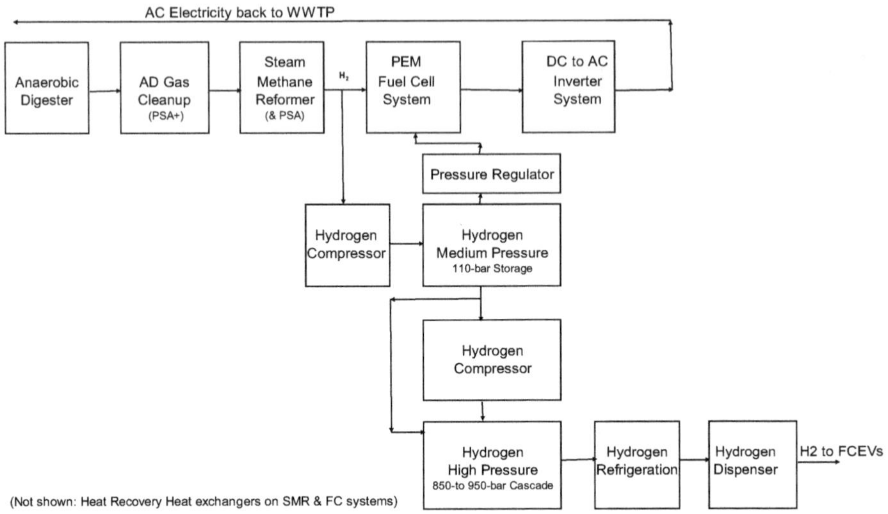

Fig. 13.1 Block diagram of the hydrogen tri-generation system at a wastewater treatment plant (WWTP) *AD* anaerobic digester; *PSA* pressure swing adsorption (gas cleanup system); *PEM* proton exchange membrane (type of low-temperature fuel cell system); *DC* direct current (electricity); *AC* alternating current

is comprised primarily of methane (CH_4, the main constituent of natural gas) and carbon dioxide (CO_2, the primary greenhouse gas).

The resulting anaerobic digester gas (ADG) is then fed into a gas cleanup system that removes the toxic gases such as hydrogen sulfide (H_2S) and siloxane, leaving a gas primarily composed of methane (CH_4) and carbon dioxide. This mixture passes into a pressure swing adsorption (PSA) or membrane filtering system that separates out most of the carbon dioxide, leaving a "biogas" composed predominantly of methane.

This methane is then fed to a steam methane reformer (SMR) that converts the methane and water to hydrogen, some carbon monoxide (CO), and carbon dioxide (CO_2):

$$3H_2O + 2CH_4 = CO + CO_2 + 7H_2 \tag{13.1}$$

A water gas shift reactor then converts most of the residual CO into more hydrogen and more CO_2:

$$H_2O + CO = H_2 + CO_2 \tag{13.2}$$

For a net reaction of

$$4H_2O + 2CH_4 = 2CO_2 + 8H_2 \tag{13.3}$$

Finally, a pressure swing adsorption (PSA) system produces a very pure stream of hydrogen suitable for the stationary fuel cell system and also suitable for the fuel cells onboard the FCEVs, and the residual CO_2 is released.

Most of this hydrogen is fed to a stationary proton exchange membrane (PEM) fuel cell system on-site that generates 1.3 MW of electrical power which displaces electricity previously purchased from the local utility to run the WWTP.

To put this 1.3-MW plant in perspective, The Eastern Research Group (ERG) estimates that, as a general "rule of thumb," it takes 4.5 million gallons per day (MGD) of waste water flow to generate 100 kW of power [1], so this 1.3-MW fuel cell plant would require a waste flow of 58.5 MGD to produce 1.3 MW of FC power. The ERG also estimates that each person generates about 100 gal of waste per day, implying that WWTP would have to collect waste from 585,000 people to provide enough fuel for a 1.3-MW plant. However, there is an existing 1.2-MW fuel cell system operating in Tulare, California, that has a population of only 59,296 [2]. But this plant in Tulare collects 4.4 MGD from domestic waste (vs. the 5.9 MGD expected from the ERG "rule of thumb") augmented by 7.1 MGD from industrial waste. So the population range needed to provide enough waste for a 1.3-MW fuel cell system varies between 60,000 and 585,000, depending on the amount of industrial waste available to process in addition to domestic waste, among other variables. In any case, there are 12.8 million people in the LA area that produce approximately 1281 MGD of waste, enough to support 28.5 MW of electrical power from just the domestic waste without any industrial or agricultural waste.

Some hydrogen from the reformer system is diverted to a compressor that raises the pressure from 13.8 bar leaving the SMR to a 500-bar medium pressure storage vessel or vessels. A second compressor raises the pressure of a small fraction of this hydrogen as high as 850–950 bar to fill 700-bar hydrogen tanks commonly used today on FCEVs.[3]

This high-pressure hydrogen intended for FCEVs has to be cooled to prevent overheating while filling the car tanks[4] that would prevent the stored hydrogen from reaching the 700-bar hydrogen capacity design point for the FCEV, so a refrigeration system is also used to cool the hydrogen before dispensing to achieve a full filling of the hydrogen tanks.

13.2 Tri-Gen System Cost Estimates

See Appendix E for a detailed assessment of component costs for this Tri-Gen system. We estimate a total system capital cost of approximately $3.56 million plus installation cost estimated at $711 thousand for a total installed cost of $4.27 million.

---

[3]The high pressure is required to reduce the volume of the hydrogen tanks on the vehicles.

[4]This mild tank heating is not a safety concern, but it would prevent the complete filling of the hydrogen tanks.

13.3 Cost Savings

Costs for the WWTP operator will be reduced by three factors:

- Reduced electricity costs;
- Reduced heating costs; and
- Income from selling hydrogen.

13.3.1 Reduced Electricity Costs

The electricity from the 1.3-MW stationary fuel cell system will displace electricity previously purchased from the local utility.[5] Over one year, the 1.3-MW fuel cell system will produce approximately 11.05 million kWh of electricity, assuming a 96 % annual availability factor. At 11.3 cents/kWh, the projected cost of industrial electricity in California for the year 2015 [3], this would save $1.25 million in avoided electricity costs. However, the SMR will require 3.62 million kWh/year of electricity and the hydrogen compressors will consume another 186,00 kWh/year of electricity, so the net savings in electricity is approximately 7.23 million kWh/year or an average of net savings of approximately $790,000 per year in avoided electricity cost.[6]

13.3.2 Reduced Heating Costs

Similarly, heat recovered from the SMR and fuel cell systems will offset heat previously generated from purchased natural gas to warm the anaerobic digester tanks. As shown in Table 13.1, we estimate that 98 MBTU/day of heat will be collected from these two systems.

The total heat recovered over one year amounts to 34,740 MBTU with a 96 % plant availability factor; with industrial natural gas selling at $5.89/MBTU in California, this amounts to a savings of approximately $204,000 per year in avoided natural gas payments to heat the digester tanks.

---

[5]WWTPs require considerable electricity; for example, the fuel cell system at the Tulare WWTP mentioned above requires 2.7 MW of electricity, so the 1.2 MW from the existing fuel cell system covers only 45 % of the required electricity to run the plant.

[6]Actual annual cost savings will vary depending on the price of industrial electricity in each region.

Table 13.1 Parameters used to calculate the heat recovered from the SMR and FC systems

		Heat recovery (MBTU/day)
Heat from SMR		
SMR eff.	75.0 %	
H2 production	1979.7 kg/day	
Power output	65,952 kWh/day	
Input energy	87,936 kWh/day	
Heat energy available	21,984 kWh/day	
Heat recovery eff.	75.0 %	
Heat energy recovery	16,488.1 kWh/day	56.3
Heat from fuel cell		
FC eff.	48 %	
Electricity output	1300 kW	
Heat available	676 kW	
Heat recovery eff.	75.0 %	
Heat energy recovery	12,168.0 kWh/day	41.5
Total heat recovery		97.8

13.3.3 Hydrogen Sales Income

We assume that hydrogen is sold to FCEV drivers either directly through a fueling station at the WWTP plant, or by shipping the hydrogen by pipeline or tanker truck to nearby hydrogen dispensing stations. The model makes the following calculations for each year of the simulation using data from the Energy Information Administration (EIA), a division of the U.S. Department of Energy. All the energy cost data in this book are obtained from the EIA's Annual Energy Outlook (AEO) for 2014 [3]:

1. *Calculate the cost per mile for gasoline in a conventional car.* For the first year of this simulation (2015), the AEO projected cost for gasoline is \$2.59/gal. With an on-the-road fuel economy of 25.6 miles/gal for a conventional gasoline internal combustion engine vehicle, the driver would pay \$2.59/25.6 = \$0.1011/mile.
2. *Calculate the cost of hydrogen.* We assume that the FCEV will pay only 90 % of the cost per mile for gasoline (to give drivers a 10 % fuel cost incentive to buy FCEVs), so the FCEV driver would only pay \$0.0909/mile.[7] We assume

[7]Thus, the hydrogen revenue used in this model is only 90 % of the revenue if the FCEV driver paid the same cost per mile as the driver of a gasoline vehicle, and the internal rates of return would be higher than reported here.

Table 13.2 Annual O&M cost for the hydrogen Tri-Gen station

Taxes and insurance	2.0 %	$71,108
Annual FC O&M	5.5 %	$40,884
Annual ADG cleanup O&M	5.8 %	$20,551
Annual WGS or SMR O&M	3.0 %	$57,678
Annual compressor O&M	4.0 %	$10,208
Annual storage O&M	0.02 %	$11
Annual dispenser O&M	0.03 %	$50
Total annual O&M costs		$200,489

that the FCEV has a fuel economy of 68.3 miles/kg of hydrogen, the fuel economy certified by two DOE National Laboratories with on-the-road testing of two Toyota fuel cell versions of their Highlander SUV [4].[8] The cost of hydrogen to reach 90 % of the cost of gasoline per mile is then $\$0.09 \times 68.3 = \$6.21$/kg of hydrogen.

3. *Calculate the amount of hydrogen consumed in FCEVs annually*. We assume an average FCEV will drive 11,876 miles per year,[9] similar to conventional cars. With a fuel economy of 68.3 miles/kg, this translates into an annual hydrogen consumption of 11,876/68.3 = 173.9 kg of hydrogen per FCEV per year. Assuming an initial design value of 60 FCEVs supported per station results in a hydrogen consumption of $173.9 \times 60 = 10{,}433$ kg/year.
4. The annual revenue for selling hydrogen at this 10 % discount is $10{,}434 \times \$6.21 = \$64{,}800$ per year.

13.3.4 Annual Operations and Maintenance (O&M) Costs

The annual O&M costs are summarized in Table 13.2. We used a 3–6 % annual O&M charge for most of the major equipment, but the DOE uses an O&M rate of only 4 % for compressors. The Tri-Gen station will incur costs of approximately $200,500 each year.

[8]This was an early version of the Toyota fuel cell stack. We anticipate that the next generation of fuel cells for commercial FCEVs will have higher efficiency, and the auto companies will be making FCEVs in smaller and lighter vehicles than SUVs, both of which will increase future FCEV fuel economy, which makes the calculations reported here conservative.

[9]This was the average US vehicle travel in 2013, see Table 7.2.

Table 13.3 WWTF cash flow (excluding Capex!) and annual savings in 2015 (first year)

Electricity avoided costs	$818,657
Reduced heating costs	$204,706
Hydrogen revenues	$64,838
O&M costs	$(200,489)
Net annual cash flow	$887,713

13.3.5 Net Annual Cost Savings

As summarized in Table 13.3, the Tri-Gen system will save the WWTF approximately $887,713 the first year. The actual savings will fluctuate each year depending on the costs of electricity, natural gas, and gasoline, which will be different across the country as projected in the EIA's Annual Energy Outlook for 2014. The average annual cash flow for California is just over $870,000 per year, including replacement of the fuel cell systems every four years.

13.4 Number of FCEVs on the Road

As described in Appendix E, the Air Resources Board of the California Environmental Protection Agency is projecting that there will be 6650 FCEVs on the road in California by 2017 and 18,465 by 2020 [5]. We are modeling 34 WWTPs in the Los Angeles area (23 in LA County and 11 in Orange County). Approximately 32.8 % of all vehicles in California are registered in these two counties,[10] so the share of FCEVs for these 34 WWTPs would be 2178 FCEVs/WWTP and 6048 FCEVs/WWTP in 2017 and 2020, respectively. For the purposes of this economic analysis, we are assuming that the Tri-Gen facilities at each WWTP are designed to support only 60 FCEVs initially,[11] or a small fraction of the likely FCEVs per WWTP after 2015.

[10]A total of 7.159 million LDVs registered in LA County and 2.437 million in Orange County or a total of 9.6 million out of 29.3 million registered vehicles in California.

[11]These systems could be designed to support more FCEVs that would increase hydrogen revenue, but at the expense of higher installed capital cost. The return on investment is larger for the smaller hydrogen refueling capacity as shown in Fig. 13.3.

13.5 Detailed Cash Flow Analysis of Tri-Gen Hydrogen Fueling Stations

The ten-year cash flow statement for a hydrogen Tri-Gen station at a WWTF in California is shown in Table 13.4. The 10-year internal rate of return (IRR) is 17.9 % on this $4.3 million investment in 2015, with a 15-year IRR of 21.7 % and a simple payback period of 4.4 years; this analysis assumes a PEM fuel cell stack life of 40,000 h, so the WWTP must replace the 1.3-MW FC system every 4 years[12] at an assumed cost of $572/kW (see Appendix E for details).

This analysis assumes that the WWTF supports 60 FCEVs initially, which is a relatively small fraction of the 2407 FCEVs projected to be on the road in these two counties by 2017. The Annual Energy Outlook (AEO-2014) is projecting relatively flat fuel costs over this period as reflected in the revenue streams shown in Fig. 13.2. Thus, the projected electricity prices fall slightly, and the hydrogen revenue (proportional to gasoline prices) and the heating costs (proportional to natural gas prices) are fairly constant over this period with the average annual positive cash flow of approximately $870,400 per year.[13] The dips in the cash flow every 4 years are due to fuel cell system replacement; since there are 8–9 FC systems to reach our 1.3 MW net capacity, these replacements could be staggered over several years to smooth out the cash flow. The project achieves positive cash flow by 2020.

The WWTP operator could also decide to build a larger hydrogen fueling station, particularly in California where many more than the average number of FCEVs is expected to be deployed in the next decade. A larger station serving more FCEVs will increase the initial Capex required and will require more operating costs, but it will also result in increased hydrogen revenues. The trade-off between Capex and the resulting return on the investment is illustrated in Fig. 13.3 for California. The 10-year IRR decreases from the 17.9 % for the design case of a station supporting 60 FCEVs to 16.1 % for a station supporting 120 FCEVs. The initial Capex increases from $4.3 million for the 60-FCEV-capacity station up to approximately $4.6 million for a 120-FCEV Tri-Gen station.[14]

---

[12]The fuel cell system would undoubtedly be a set of seven 100-kW stacks, so these replacements could be staggered over the years.

[13]Not shown in Fig. 13.4 are the annual O&M costs of approximately $142,000 each year, although these costs are included in the calculation of total net cash flow line of approximately $768,000 per year.

[14]Approximately 70 % of the Tri-Gen system costs are associated with the gas cleanup fuel cell system, and other components are not associated with FCEV fueling, so system costs do not increase significantly with increased numbers of FCEVs supported by the facility.

Table 13.4 Cash flow for a Tri-Gen WWTP with a capacity to support 60 FCEVs in California

Year⟹		2015	2016	2017	2018	2019	2020	2021	2022	2023	2024
Capex		$(3,555,394)	$–	$–	$–	$(407,909.7)	$–	$–	$–	$(407,909.7)	$–
Installation		$(711,078.73)	$–	$–	$–	$(81,581.9)	$–	$–	$–	$(81,581.9)	$–
Taxes and insurance		$(71,107.87)	$(71,108)	$(71,108)	$(71,108)	$(71,108)	$(71,108)	$(71,108)	$(71,108)	$(71,108)	$(71,108)
Annual FC O&M	$(1,63,535.43)	$(40,883.86)	$(40,884)	$(40,884)	$(40,884)	$(40,884)	$(40,884)	$(40,884)	$(40,884)	$(40,884)	$(40,884)
Annual ADG cleanup O&M		$(20,551.27)	$(20,551)	$(20,551)	$(20,551)	$(20,551)	$(20,551)	$(20,551)	$(20,551)	$(20,551)	$(20,551)
Annual PSA cleanup O&M			$–	$–	$–	$–	$–	$–	$–	$–	$–
Annual WGS or SMR O&M		$(57,677.63)	$(57,678)	$(57,678)	$(57,678)	$(57,678)	$(57,678)	$(57,678)	$(57,678)	$(57,678)	$(57,678)
Annual compressor O&M		$(10,207.58)	$(10,208)	$(10,208)	$(10,208)	$(10,208)	$(10,208)	$(10,208)	$(10,208)	$(10,208)	$(10,208)
Annual storage O&M		$(11.02)	$(11)	$(11)	$(11)	$(11)	$(11)	$(11)	$(11)	$(11)	$(11)
Annual dispenser O&M		$(49.50)	$(50)	$(50)	$(50)	$(50)	$(50)	$(50)	$(50)	$(50)	$(50)
Gross avoided electricity costs		$1,249,834	$1,232,691	$1,216,952	$1,206,889	$1,197,625	$1,184,071	$1,170,061	$1,158,845	$1,145,686	$1,135,962

(continued)

Table 13.4 (continued)

Year⟹		2015	2016	2017	2018	2019	2020	2021	2022	2023	2024
Natural gas costs		$–	$629,760	$659,145	$708,579	$730,619	$723,297	$734,921	$740,853	$751,085	$765,531
Electricity for SMR		$410,064	$404,439	$399,275	$395,973	$392,934	$388,487	$383,890	$380,211	$375,893	$372,703
Electricity for H2 compression		$21,114	$20,824	$20,558	$20,388	$20,232	$20,003	$19,766	$19,577	$19,354	$19,190
Net avoided electricity costs		$818,657	$807,428	$797,119	$790,527	$784,460	$775,581	$766,405	$759,058	$750,439	$744,069
Avoided heating costs		$204,706	$205,685	$215,282	$231,428	$238,626	$236,234	$240,031	$241,968	$245,310	$250,028
Hydrogen revenue		$64,838	$62,732	$61,857	$60,626	$59,409	$61,824	$59,904	$59,307	$57,783	$57,221
Cash flow		$(3,378,760)	$875,355.36	$873,768.46	$882,091.54	$392,514.30	$873,150.81	$865,851.20	$859,844.75	$363,551.32	$850,829.94
Cumulative cash flow		$(3,378,760)	$(2,503,404)	$(1,629,636)	$(747,544)	$(355,030)	$518,121	$1,383,972	$2,243,817	$2,607,368	$3,458,198
10-year IRR:	17.9 %										
15-year IRR	21.7 %										
Payback period (years)	4.4-years										

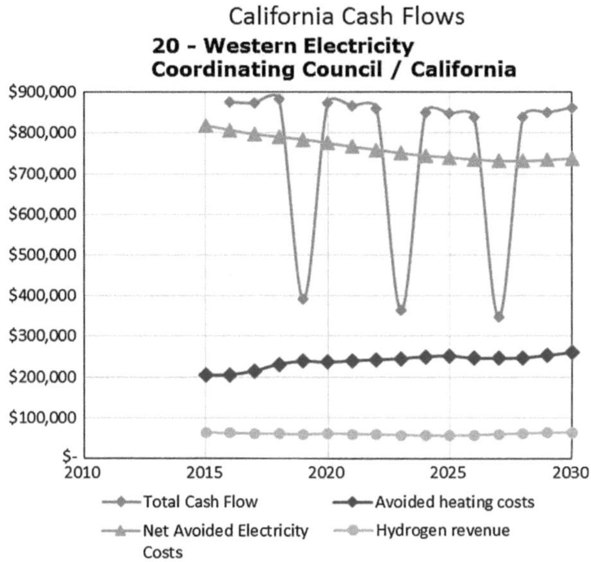

Fig. 13.2 Cash flow components and fossil fuel prices for California case (Negative cash flow of ≈ \$142,000 per year for O&M costs not shown) (Four-year dips are due to periodic fuel cell replacements)

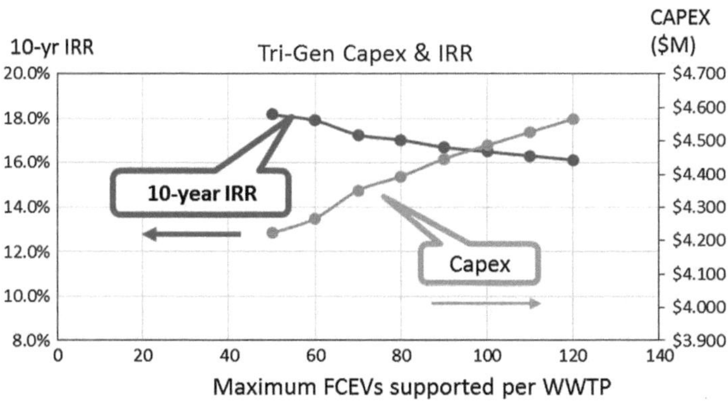

Fig. 13.3 Trade-off between Capex and return on investment versus the capacity of the hydrogen fueling station (number of FCEVs supported) for California

13.6 Impact of Lower Gasoline Prices

With the rapid drop in gasoline prices in the 2014–2015 time period, one might be concerned that the IRR would fall if hydrogen prices had to be reduced to match the lower gasoline prices. Since the cash flow contribution of hydrogen sales is relatively small, reduced gasoline prices (assuming that hydrogen is sold at 10 %

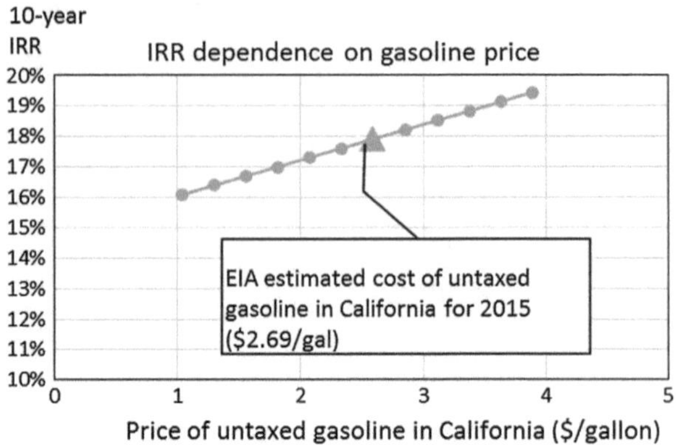

Fig. 13.4 The 10-year project IRR as a function of untaxed retail gasoline in California

discount to gasoline on a cost per mile basis) do not have a major impact on return on investment as shown in Fig. 13.4. The EIA was projecting a cost of \$2.59/gal for untaxed gasoline in California in 2015 which would yield a project 10-year IRR of 17.9 %. If untaxed gasoline prices fell to \$1/gal, then the IRR would still be a healthy 16 %, and if gasoline prices should increase to \$3.60/gal, then the IRR would rise to 19.1 %.

13.7 Tri-Gen WWTF Economic Performance in Other Areas of the Country

The performance of the WWTF hydrogen stations will depend primarily on the price of industrial electricity in each region, the number of FCEVs supported by the facility, and, to a lesser degree, by the local industrial fuel prices for natural gas and gasoline.

The IRRs and payback periods are listed in Table 13.5 for the 22 electricity regions used by the EIA.[15] The "FCEV multiplier" is the factor that multiples the base model national average number of vehicles over the 2015–2029 time period. We assumed a multiplier of one for most of these regions, 225 for California[16] (see Appendix E), and two for Region 8 (upstate New York), since there have been discussions about placing hydrogen stations and FCEVs in New York that might push it above the national average of FCEV density.

Remarkably, there are reasonable IRRs in the Northeast even without a FCEV multiplier greater than one, with a WWTP Tri-Gen facility on Long Island earning a healthy 18.3 % 10-year return, 22.5 % 15-year return, and 14.3–19.3 % for

[15]See Fig. 13.5 for a map of these regions.

[16]This 225 multiplier factor accounts for the much higher density of FCEVs in southern California than the national average in the 2015–2020 time period.

Table 13.5 Performance of hydrogen Tri-Gen stations in other regions of the USA

	FCEV Multiplier	10-year IRR (%)	15-Year IRR (%)	Payback Period (years)
01—Texas	1	−2.5	5.6	9.9
02—Florida	1	8.8	14.4	6.0
03—Midwest Reliability Council/East	1	2.0	8.7	8.4
04—Midwest Reliability Council/West	1	−1.8	5.9	9.6
05—Northeast Power Coordinating Council/New England	1	17.7	22.1	4.6
06—Northeast Power Coordinating Council/NYC-Westchester	1	13.5	18.7	5.2
07—Northeast Power Coordinating Council/Long Island	1	18.3	22.5	4.4
08—Northeast Power Coordinating Council/Upstate New York	1	14.3	19.3	5.0
08—Northeast Power Coordinating Council/Upstate New York	2	14.7	19.5	4.9
09—Reliability First Corporation/East	1	0.7	8.2	8.8
10—Reliability First Corporation/Michigan	1	3.8	10.3	7.4
11—Reliability First Corporation/West	1	2.7	9.2	8.0
12—SERC Reliability Corporation/Delta	1	1.0	8.2	8.7
13—SERC Reliability Corporation/Gateway	1	−2.8	5.2	10.0
14—SERC Reliability Corporation/Southeastern	1	−0.6	6.7	9.2
15—SERC Reliability Corporation/Central	1	−3.2	4.7	10.1
16—SERC Reliability Corporation/Virginia-Carolina	1	2.3	9.1	8.3
17—Southwest Power Pool/North	1	1.8	8.6	8.4
18—Southwest Power Pool/South	1	−6.5	2.6	12.3
19—Western Electricity Coordinating Council/Southwest	1	0.8	8.0	8.8
20—Western Electricity Coordinating Council/California	225	16.5	20.6	4.6
20—Western Electricity Coordinating Council/California	225	17.9	21.7	4.4
21—Western Electricity Coordinating Council/Northwest Power Pool Area	1	−8.8	0.8	13.5
22—Western Electricity Coordinating Council/Rockies	1	−0.03	7.4	9.0

Fig. 13.5 Map of 22 EIA utility regions in the USA

Region #8 (upstate New York). Two rows are shown for Regions 8 and 20: The first row has a unity FCEV multiplier to allow comparisons with the other regions, and the second row has a factor of 2 multipliers on the base model number of FCEVs in upstate New York, showing the doubling the small number of FCEVs increases the 10-year IRR modestly from 14.3 to 14.7 %. In California, the IRR increases from 16.5 % for a unity multiplier to 17.9 % with the expected multiplier of 225 to one based on anticipated California FCEV density and hydrogen station deployments.[17]

[17]A multiplier does not need to be his large; a multiplier of 60 is sufficient to assure at least 60 FCEVs in the region of each WWTP starting in 2015. In other words, the IRRs reported here would still be accurate if the FCEV density were only increased by a factor of 60 above the national average; thus, the projected 225 times multiplier projected by the California Air Resources Board could be reduced by a factor of approximately 3.75 and the Tri-Gen facility would still earn a 17.9 % 10-year IRR.

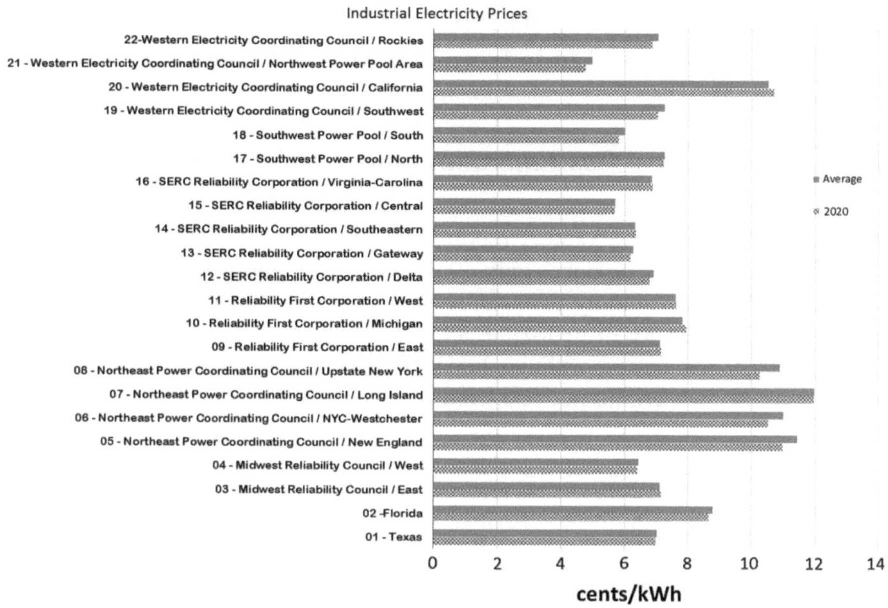

Fig. 13.6 Estimated costs of industrial electricity in the 22 EIA regions; the *solid bars* show the average industrial prices over the 2015–2030 time period from AEO-2014, while the *cross-hatched bars* show the electricity prices projected for the year 2020. Regions with the highest industrial electricity prices provide the highest returns for a hydrogen Tri-Gen system since the stationary fuel cell systems displace purchased electricity, which is the primary positive cash flow for the project

The regions in the Northeast all have positive IRRs even without an increase in FCEVs over the low model national average projections. This is due primarily to the industrial electricity prices projected by the DOE's EIA being high in these regions as shown in Fig. 13.6, which shows the average industrial electricity prices over the 2015–2030 time frame and also the prices in 2020 (crosshatched bars). From Table 13.3, the avoided electricity costs account for approximately 96 % of the positive annual WWTP cash flow, so electricity prices dominate the IRR calculation.[18] As shown in Fig. 13.6, the Northeast and California have the highest projected industrial electricity prices and are therefore the best locations for hydrogen Tri-Gen fueling stations at WWTPs in the early years (all locations will be more profitable once there are larger numbers of FCEVs on the road).

[18]In a more conventional hydrogen production plant where a reformer converts natural gas to electricity, the appropriate figure of merit is the "spark spread," the difference between the price of electricity and natural gas, both in the same units such as $/MBTU. But in this Tri-Gen system, we are converting waste to hydrogen, so the local price of natural gas has only a minor impact through the heat savings; the input fuel (solid waste) is free.

13.8 Greenhouse Gas Emissions and Oil Consumption with Hydrogen Made from Wastewater Treatment Plant Waste

13.8.1 Greenhouse Gas Emissions with WWTP Hydrogen

The Argonne National Laboratory reports that a fuel cell EV running on hydrogen made from landfill gas produces only 40 g/mile of CO_2 equivalent GHG emissions with a FCEV fuel economy of 60 miles/gal of gasoline equivalent [6] which is equal to 57.37 miles/kg of hydrogen; adjusting for our assumption of 68.3 miles/kg for a FCEV reduces the estimated GHG emissions to 35.1 g/mile, compared to a minimum of 56.9 g/mile in 2100 from the mix of hydrogen sources considered in Chap. 8.

13.8.2 Hydrogen Production Potential from Biogas

Saur and Milbrandt [7] of the National Renewable Energy Laboratory analyzed the amount of available hydrogen that could be generated from wastewater treatment plants (WWTPs) and landfills. Their analysis showed that these two sources could produce over 1.1 million tonnes of hydrogen per year, enough to support 6.6 million FCEVs, which is approximately 2.6 % of the current light-duty vehicle (LDV) fleet of 253.6 million vehicles as shown in Table 13.6. These estimates are conservative since the NREL data for WWTPs are based on only the 18,000 of the 35,000 WWTPs in the USA that provided data on wastewater flow, which are necessary to estimate the hydrogen potential from each plant. Their data do not include WWTPs that already utilize at least some of their biogas to either produce electricity or heat their anaerobic digester tanks that are essential to generating the biogas.

We conclude that hydrogen made from wastewater treatment plants and landfill gases can support only 2.6 % of all US light-duty vehicles based on the limited WWTP data available, so this will not be a major source of hydrogen in the long run. However, given the very favorable economics described above, we conclude that these Tri-Gen systems may be an excellent economic opportunity to install a set of zero-carbon hydrogen fueling stations across the country that would bring a healthy return on investment for investors.

Table 13.6 Renewable hydrogen available from waste sources

	Hydrogen available (ktonnes/year)	#FCEVs supported	Percentage of US LDVs
WWTP	509	2,927,307	1.2
Landfills	648	3,726,709	1.5
Totals	1157	6,654,017	2.6

13.8.3 Oil Consumption with WWTP Hydrogen

Oil consumption will also be reduced slightly by adding the hydrogen made from waste as described in the previous section. The Argonne National Laboratory analysis calculated an oil consumption rate of only 10 BTU/mile for FCEVs using the hydrogen from waste [7], which is less than the 79 BTU/mile calculated for 2100 with the hydrogen production mix used in Chap. 8.

References

1. “Opportunities for and Benefits of combined heat and power at wastewater treatment plants,” by the Eastern Research Group(ERG) for the U.S. Environmental Protection Agency, December 2006.
2. “Fuel Cell system turns waste into electricity at the Tulare wastewater treatment plant,” by Fuel Cells2000, available at www.fuelcells.org/uploads/tularecasestudy.pdf
3. The Annual Energy Outlook for 2014 by the Energy Information Administration of the U.S. Department of Energy, available at: http://www.eia.gov/oiaf/aeo/tablebrowser/
4. K. Wipke, D. Anton & S. Sprik, “Evaluation of range estimates for Toyota FCV-adv under open road driving conditions,” National Renewable Energy and Savannah River National Laboratories, SRNS-STI-2009-00446, August, 2009.
5. “An annual evaluation of fuel cell vehicle deployment and hydrogen fuel station network development,” Air Resources Board of the California Environmental Protection Agency, June 2014, available at http://www.arb.ca.gov/msprog/zevprog/ab8/ab8_report_final_june2014.pdf
6. “Expanding the use of biogas,” presented by Fred Joseck of the DOE Fuel Cell Technologies Office to “The Innovative Energy & environmental Technology Workshop, Power generation and sustainable biomass in New Jersey”, Bordertown, New Jersey, 5/7/2012. Available at: http://ecocomplex.rutgers.edu/IEETW_Fred%20Josek_RutgersWorkshop.pdf
7. G. Saur & A. Milbrandt, “Renewable Hydrogen Potential from Biomass in the United States,” National Renewable Energy Laboratory Report NREL/TP-5400-60283, available at: http://www.nrel.gov/docs/fy14osti/60283.pdf

Chapter 14
Automobile Companies on FCEVs

Abstract Most major automobile companies have been developing FCEVs over the last two decades. This chapter includes some quotations from various auto executives regarding the efficacy of FCEVs and BEVs.

14.1 DaimlerChrysler

> "*The fuel cell is the most promising option for the future. We are determined to be the first to bring it to market.*"—Juergen Hubbert, DaimlerChrysler

14.2 Ford

> "*I believe fuel cells could end the 100-year reign of the internal combustion engine. In 25 years, fuel cells could be the predominate automotive power source. It will be a winning situation all around – customers will get an efficient power source, communities will get zero emission transportation and automakers will get a major business opportunity.*" Bill Ford, Chairman, Ford Motor Company, International Auto Show, January 2000

14.3 GM

> "*Of all the technologies, the fuel cell car seems to be the most promising, it has a good chance of becoming the next mass market car.*" Byron McCormick, co-director of General Motor's Global Alternative Propulsion Center

C.E. (Sandy) Thomas, *Sustainable Transportation Options for the 21st Century and Beyond*, DOI 10.1007/978-3-319-16832-6_14

14.4 Honda Motor

"*Fuel cell vehicles will probably overtake gasoline-powered cars in the next 20 to 30 years.*" Takeo Fukui, managing director, Research and Development, Honda Motor Co., Bloomberg News, June 5, 1999

"*Fuel cell vehicles, which run on hydrogen and emit no CO_2 or other harmful emissions, may be the key to securing a bright future for the automobile in society,*" Sachito Fujimoto, FCX Clarity Project Leader

On consumer demand for battery electric vehicles (BEVs) "*It's questionable whether consumers will accept the annoyances of limited driving range and having to spend time charging them,*" Tomohiko Kawanabe, Honda's President of Research and Development

14.5 Toyota

"The automobile industry can contribute to the sustainable growth of earth itself; At Toyota, we are looking out 50 years and even more decades into the future. I do believe that [the] fuel-cell vehicle is the ultimate environmentally friendly car. But the point is not just to introduce it as an eco-friendly car with good mileage. I wanted it to be fun to drive and interesting as a car."—Akio Toyoda, Toyota Motor's president 12/17/2014; available at: http://www.businessweek.com/articles/2014-12-17/toyota-embraces-fuel-cell-cars-for-post-gasoline-future

We are at a turning point in automotive history. A turning point where people will embrace a new, environmentally friendly car that is a pleasure to drive.

A turning point where a four-door sedan can travel 300 miles on a single tank of hydrogen, can be refuelled in under five minutes and emit only water vapour. A turning point that represents many years and countless hours of work by our team to create a car that redefines the industry.

All of us at Toyota believe in a future that will be safer, greener and easier for everyone. We imagined a world filled with vehicles that would diminish our dependence on oil and reduce harm to the environment. It was a bold, but inspiring goal and today it is a reality.

"Our fuel cell vehicle runs on hydrogen that can be made from virtually anything, even garbage. It has a fuel cell that creates enough electricity to power a house for about a week." —Akio Toyoda, President of Toyota, Dec 22, 2014 Available at: http://www.autocarpro.in/news-international/hydrogen-powered-2015-toyota-mirai-price-info-specs-7151

"*Because of its shortcomings - driving range, cost and recharging time - the electric vehicle is not a viable replacement for most conventional cars. We need something entirely new.*" Vice Chairman Takeshi Uchiyamada, who said he believes that fuel-cell vehicles hold far more promise than battery electric cars. Tokyo, September 24, 2012 http://www.cnbc.com/id/100430620

"*Hydrogen is a real solution into the future that solves our carbon problem, as well,*" Jim Lentz, CEO of Toyota North America Sept. 27, 2013, Northwood University International Auto Show in Midland, Michigan

"Our unique hybrid history and experience have proven invaluable for the next big leap. Back in 2010, we promised our first fuel cell car for 2015 and we are fully on track to honour our commitment. **Fuel cell is a technology that can secure our concept of personal mobility. That's because fuel cells combine the strengths of EVs (electric vehicles) and hybrids, with those of conventional cars. That means zero emissions – they only emit water vapour – and full usability; refuelling only takes about three minutes.**" Karl Schlicht, Executive Vice President of Toyota Motor Europe, April 16, 2014

On Toyota's research priorities: "*One prime example is what I call the NEXT BIG THING…a zero-emission hydrogen fuel cell sedan we'll launch to the public in California next summer. It's an electric* ***car****…but instead of carrying around a huge, expensive battery, it carries compressed hydrogen gas…and a relatively small battery. As needed and on-demand: a fuel-cell stack mixes two elements abundant in nature …hydrogen with oxygen…to produce electricity that powers the* ***vehicle*** *for about 300 miles on a single fill-up which takes about three minutes. So it has the range of today's conventional cars without lengthy re-charging… while emitting only harmless water vapor…the best of BOTH worlds. We've reduced the cost of the fuel cell powertrain by 95 %…and we're confident we can reduce the cost further.*" Bob Carter, Senior VP, Automotive Operations, Toyota Motor Sales, from the J.P. Morgan Auto Conference, New York City, August 12, 2014

"*Plug-in hybrids make most sense if you live in a city, but people who live in cities usually have to park on the street, so where are they supposed to charge a car? We believe the fuel cell is a much better technology because it allows you to refuel your car with hydrogen in the same way you refuel it with petrol today. We can't argue with that. Electric vehicles are a step above hybrids, and we think hydrogen cars can be even better than both. Now we just wait for Lexus to make a luxury hydrogen car.*" Alain Uyttenhoven, VP of Lexus Europe

Chapter 15
Conclusions

Abstract We conclude that the hydrogen-powered fuel cell electric vehicle (FCEV) is the best and only option to simultaneously reduce greenhouse gas (GHG) emissions, local air pollution, and the consumption of fossil fuels such as petroleum and natural gas. The primary advantage of the FCEV is that it is the *only* alternative vehicle option that can achieve the societal goal of reducing GHG emissions by 80 % below 1990 levels. But the FCEV also produces the lowest cost urban air pollution, the lowest fueling infrastructure cost per vehicle, and will reach our goal of cutting petroleum consumption to the point where all of our oil could be provided by countries on the American continent (excluding Venezuela!) in an emergency nearly as soon as than any other option.

15.1 Transportation Conclusions

15.1.1 Greenhouse Gas Emissions

In the short term (2015–2020), battery EVs (BEVs) will reduce GHGs by 3.6 % per vehicle compared to gasoline HEVs, gasoline plug-in HEVs in the USA will decrease GHG emissions by 15.7 %, while hydrogen-powered fuel cell electric vehicles (FCEVs) will reduce GHGs by 22.2 % even when all hydrogen is made from natural gas in the near term;

In the longer term (over the twenty-first century), the goal of the climate change community is to reduce GHG emissions to 80 % below 1990 levels, preferably by 2050. Our modeling (Fig. 15.1) shows that gasoline-powered hybrid electric vehicles (HEVs) like the Toyota Prius would reduce GHGs by approximately 60 % below 1990 levels by the end of the twenty-first century. Gasoline-powered plug-in EVs (PHEVs) like the Chevy Volt would reduce GHGs by 55 % below 1990 levels. Ethanol-powered PHEVs could reduce GHGs by 66.4 % below 1990 levels by 2100, while BEVs could reduce GHG emissions by approximately 70 % below 1990 levels (limited primarily by a low percentage of vehicles that could be powered only by batteries).

C.E. (Sandy) Thomas, *Sustainable Transportation Options for the 21st Century and Beyond*, DOI 10.1007/978-3-319-16832-6_15

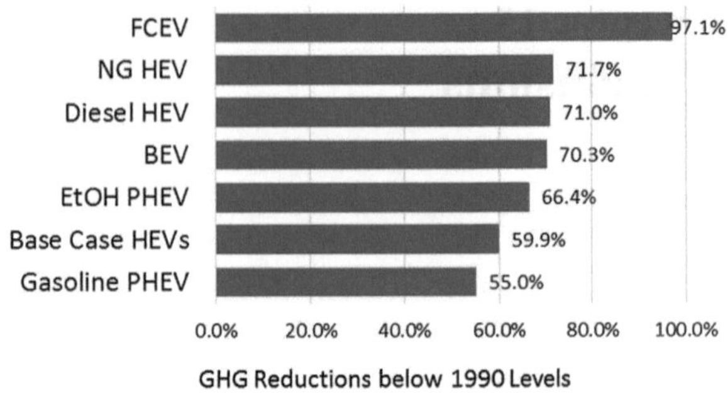

Fig. 15.1 Estimated reduction in greenhouse gas (GHG) emissions by 2100 as a percentage below 1990 levels (societal goal is an 80 % reduction preferably by 2050)

The hydrogen-powered fuel cell EV is the only option that can reach the 80 % reduction goal, achieving that milestone by approximately 2060 with the FCEV market penetration levels and mixture of "green" hydrogen production systems assumed in our model, achieving approximately a 97.1 % reduction below 1990 levels by the end of the century according to our model.

15.1.2 Petroleum Consumption

In the short term, a natural gas HEV provides the greatest reduction in oil consumption, reducing oil use approximately 98 % below a gasoline HEV; a BEV is second best *on a per vehicle basis*, cutting oil consumption approximately 92 % according to the Argonne GREET model; each FCEV cuts oil consumption by 72 % compared to a gasoline HEV, while an ethanol HEV reduces oil consumption by 65 % compared to an HEV.

In the longer term, taking into account the likely market potential of alternate vehicles, the FCEV is the best alternative,[1] reducing light-duty vehicle oil consumption to the point where the USA could supply all oil needs from the non-OPEC nations by 2043 and from nations on the American continent by 2047. The BEV scenario could achieve this non-OPEC goal about three years earlier with our model parameters, while all other vehicle options would achieve those two petroleum reduction goals by the 2045 to 2051 time period as shown in Table 9.3.

[1]Other than a natural gas HEV, which would reduce petroleum consumption more, but would increase urban air pollution costs and could not approach our GHG reduction goal, we conclude that natural gas vehicles would not be acceptable due to increased urban air costs and the lack of a viable pathway to achieve GHG reduction goals.

15.1.3 Fuel Infrastructure Cost

While hydrogen fueling stations cost more than battery charging stations, each hydrogen station can serve hundreds to thousands of vehicles, while battery charging outlets can serve only one or two BEVs per day due to the long charging times required. Therefore, the cost of fueling infrastructure for a FCEV is much less than the cost per battery electric vehicle. As shown in Chap. 7, *the low estimate for the BEV charging infrastructure per vehicle ($7252/BEV) costs 5 times more than the high FCEV estimate ($1,346/FCEV).*

The highly respected consulting firm, McKinsey & Company, estimated that installing a charging network for BEVs in Europe would cost *five times more* than installing an EU hydrogen infrastructure [1].

15.1.4 Urban Air Pollution

As shown in Chap. 10, the hydrogen-powered FCEV produces the lowest cost urban air pollution. A BEV generates urban air pollution that is estimated to cost 4.3 times more than FCEV emissions, due primarily to the SO_x emissions from electrical plants within the urban airshed used to recharge EV batteries that contributes to the formation of smog in the presence of sunlight; a diesel HEV generates urban pollution that is 3.2 times more costly than a FCEV; a natural gas HEV pollution is 1.9 times more costly; and a gasoline HEV generates local pollution that is 4.1 times more costly than that from a FCEV, as shown in Fig. 15.2.

15.1.5 Alternative Vehicle Cost

Our cost analysis (Chap. 5) including the impact of mass compounding suggests that the power train of a BEV with 500 km (311 miles) range would cost

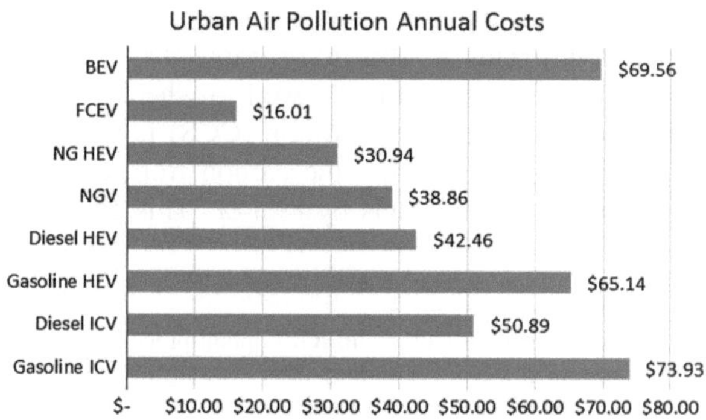

Fig. 15.2 Estimated annual air pollution costs per vehicle (due primarily to health-related costs of urban pollution)

approximately 3.5 times the cost of a hydrogen FCEV power train with that range. Two outside sources (MIT and McKinsey & Company) apparently did not consider the impact of mass compounding, so their cost differences are smaller. For example, MIT estimated that the incremental cost of a BEV with 200 miles range compared to a conventional vehicle would be $10,200, while the incremental cost of a FCEV with 390 miles range would be only $3600. McKinsey estimated even smaller differences (10 %) for the cost of BEV and FCEV J-segment SUV-like vehicles, but apparently did not include the effects of mass compounding which makes BEVs much more expensive for longer range vehicles.

15.2 Electricity Generation

While this book has focused primarily on the transportation sector, the best alternative vehicle alternative–the hydrogen-powered FCEV–would enable significant reductions in greenhouse gas (GHG) emissions from electricity generation via the use of hydrogen as an energy carrier.

One very clean source of hydrogen would be from coal at an integrated gasification combined cycle (IGCC) power plant. Rather than burning coal to create steam as in conventional coal power plants, the IGCC plant first turns the coal into a "synthesis gas" or "syngas" made up primarily of hydrogen and carbon monoxide (CO). The CO in the syngas can be turned into more hydrogen in a water gas shift reactor: $CO + H_2O = CO_2 + H_2$. The hydrogen can be used to power FCEVs or it can be used to produce electricity in a gas turbine, or preferably a stationary fuel cell system. The CO_2 can be captured and stored underground more readily than trying to capture the more dilute CO_2 in the smokestack of conventional coal plants.[2]

15.2.1 Clean Coal Hydrogen and Electricity Generation

While some people may be skeptical about the claims of "clean coal," the GHGs from an IGCC plant are actually much less than those from conventional power generation plants as shown in Table 15.1, taken from the Argonne National Laboratory 2012 "GREET" "well-to-wheels" GHG calculation program. While carbon capture and sequestration (CCS) is not 100 % perfect, Table 15.1 shows that capturing and storing the carbon from a coal IGCC plant are estimated by Argonne to cut the GHGs by 89.1 %. This coal IGCC level with CCS (85.3 g of

[2]The concentration of CO_2 in an IGCC plant precombustion gas stream is much higher than the concentration in a boiler chimney after burning the coal, making separation and capture of the CO_2 much easier for the IGCC system.

Table 15.1 Estimated greenhouse gases (GHGs) from electrical power plants in grams of CO_2-equivalent/kWh, with and without carbon capture and storage (CCS) and percentage reductions of plants with CCS below NG combined cycle plants without CCS

	No CCS	With CCS	CCS percentage reduction (%)	Percentage below NG CC (%)
Oil boiler	946.1	72.5	−92.3	−81.1
NG boiler	678.8	50.6	−92.5	−86.8
NG single cycle	656.0	50.6	−92.3	−86.8
NG combined cycle	384.3	50.7	−86.8	−86.8
Coal boiler	1048.5	85.3	−91.9	−77.8
Coal IGCC	781.2	85.3	−89.1	−77.8

CO_2-equivalent/kWh) is 77.8 % less than the GHGs from today's lowest emitting electrical power plant without CCS: the natural gas combined cycle plant (NG CC) at 384 g/kWh.

Table 15.1 assumes that the hydrogen is burned in a gas turbine. Another option with potentially higher efficiency (and thus lower GHGs) is to use the hydrogen to power a stationary fuel cell system. The GREET model assumes that the coal IGCC system is 46.7 % efficient. The DOE's National Energy Technology Laboratory (NETL) estimates that the efficiency of an IGCC driving a solid oxide fuel cell (SOFC) could be 61.3 % by 2020 and up to 64 % in the long run [2].

15.2.2 Hydrogen as Storage Pathway to Enable More Intermittent Renewable Utilization

Another benefit of producing hydrogen with a coal IGGC system is to enable greater use of intermittent renewables such as wind or solar. Thus, hydrogen can be stored during sunny and windy days and used to generate electricity through a stationary fuel cell during cloudy and calm days. This eliminates or minimizes the need to have backup fossil fuel-powered electrical generators on standby when renewables are not available and thereby reduces the net GHG emissions.

Large-scale hydrogen storage can also provide grid electricity during planned and unplanned power plant maintenance periods and alleviate the impact of some power outages.

15.3 Final Summary Conclusions

1. Hydrogen is required to support FCEVs, the *only* alternative vehicle that can plausibly achieve our societal goal of cutting GHG emissions by 80 % below 1990 levels.

2. Hydrogen made from coal by IGCC (or stationary fuel cell system) generation with carbon capture and storage or sequestration can enable the continued utilization of coal, the most plentiful fossil fuel in many developed nations in an environmentally beneficial manner, and
3. Hydrogen made from coal and stored for later use can also enable increased utilization of intermittent renewables.

References

1. McKinsey & Company, "A Portfolio of Power-Train options for Europe: a fact-based analysis: the role of battery electric vehicles, plug-in hybrids, and fuel cell electric vehicles," undated. Available at :http://ec.europa.eu/research/fch/pdf/a_portfolio_of_power_trains_for_europe_a_fact_based__analysis.pdf.
2. Assessment of the Distributed Generation Market Potential for Solid oxide Fuel Cells," DOE/NETL 342/093013, September 29, 2013. Available at: http://www.netl.doe.gov/File%20 Library/Research/Energy%20Analysis/Publications/FINAL_DG_SOFC.pdf.

Appendix A
Range Limitations of Battery Electric Vehicles

One of the key factors limiting the widespread adoption of battery electric vehicles (BEVs) is their limited range. While in principle the range of any electric vehicle could be extended by adding more batteries, in reality there are real-world limits to the range of a BEV.

The practical range of a BEV is limited by a process called "mass compounding": For every kg of battery mass added to increase range, the size and weight of other BEV components must also be increased to maintain the performance and safety of the vehicle; adding more batteries requires at a minimum:

- The mechanical system holding the battery bank must be enlarged to properly support those extra batteries.
- The electric motor and controller size must be increased to provide adequate acceleration for the BEV to safely enter a high-speed roadway or to safely pass another vehicle.
- The brakes must be enlarged to safely stop the BEV under all conditions.
- The entire vehicle structure should be strengthened to adequately protect passengers in a crash.
- The suspension system should be enlarged to provide a comfortable ride with the extra mass.
- Finally, still more batteries must be added as a result of these extra mass components listed above to achieve the desired range, and the process steps outlined above must be repeated which lead to still larger components in a nonlinear feedback process.

C.E. (Sandy) Thomas, *Sustainable Transportation Options for the 21st Century and Beyond*, DOI 10.1007/978-3-319-16832-6

Battery Specifications

We consider two types of batteries: High-energy batteries required for BEVs to maximize range, and high-power batteries required for peak power augmentation on fuel cell electric vehicles (FCEVs)

High-Energy Batteries for BEVs

The best-selling BEV in the USA, the Nissan Leaf, uses advanced lithium-ion batteries. The main characteristics of the Leaf Li-ion battery are summarized in Table A.1. The calculated useful specific energy for this battery is 54.4 Wh/kg. Nissan claims a much higher specific energy of 140 Wh/kg, but the 54.4 Wh/kg is calculated directly from the Nissan battery data [A.1], which lists a useful energy of 16 kWh and a total battery pack mass of 293.9 kg.[1]

The Nissan spec sheet also claims a specific power of 2.5 kW/kg, compared to the calculated specific power of 306 W/kg or 0.306 kW/kg. Thus, the Nissan claim is high by a factor of 8.2. Put it another way, if the specific power was really 2.5 kW/kg, then the 293.9-kg battery would have a power capacity of 734.8 kW, far above the stated power level of "above 90 kW."[2]

High-Power Batteries for FCEVs

FCEVs typically include a battery to provide peak power acceleration and also to improve efficiency by storing the energy generated with regenerative braking. However, the peak power battery for a FCEV does not need to be as heavy as a BEV battery, since its main function is to provide power, not energy.[3] As shown in Fig. A.1, most batteries can be tailored to provide high power at the sacrifice of less energy storage. By reducing the battery cell plate thickness, the battery can respond more quickly and provide more power, moving to the right in Fig. A.1, but

[1]The battery does store 24 kWh of energy, but not all energy can be used from any battery without seriously shortening its lifetime or destroying the battery and creating the havoc of running out of energy in some remote location. The useful energy assuming 67 % depth of discharge (DOD) is 16 kWh. But even if you did completely discharge the battery, the specific energy would still be only 24 kWh/293.9 kg = 0.08166 or 81.66 Wh/kg, still well short of the claimed specific energy of 140 Wh/kg.

[2]The most likely explanation for these wide differences in claimed and calculated parameters may be that Nissan is listing the attributes of individual battery *cells or modules*, and not the full battery pack; for example, the individual cells have a calculated specific energy of 157 Wh/kg and each module (4 cells) has a specific energy of 132 Wh/kg, and the average of these two values is 145 Wh/kg, close to the claimed 140 Wh/kg.

[3]The hydrogen stored onboard a FCEV provides the energy required for long driving range, eliminating the need for a heavy energy storage battery bank.

Table A.1 Key parameters of the Nissan Leaf Li-ion battery pack

		Specific power (W/kg)	Nominal specific energy (Wh/kg)	Useful specific energy (Wh/kg)	Nominal energy density (Wh/l)	Useful energy density (Wh/l)
Battery power (kW)	90	306.2				
Nominal energy (kWh)	24		81.7		485.6	
Useful energy (kWh)[a]	16			54.4		323.7
Battery mass (kg)	293.9					
Battery volume (l)[b]	49.4					

[a]24 kWh stored; 16 kWh useful at 67 % DOD
[b]1.57 m × 0.1188 m × 0.265 m

at the expense of reduced energy storage capacity. The Leaf Li-ion battery (54.4 WH/kg useful specific energy and 306 W/kg useful specific power) is shown on this chart along with the FCEV battery with higher specific power (400 W/kg) at the expense of lower specific energy (≈22 Wh/kg). Thus, the FCEV battery is closer to the battery used in a hybrid electric vehicle (HEV) than to a BEV ("EV goal" in Fig. A.1).

Mass Compounding

Malen and Reddy measured the mass compounding effects of 32 late model (2002–2007) vehicles [A.2]. They found that adding a load such as 100 kg of batteries will require an additional 59.8 kg for 12 vehicle subsystems such as

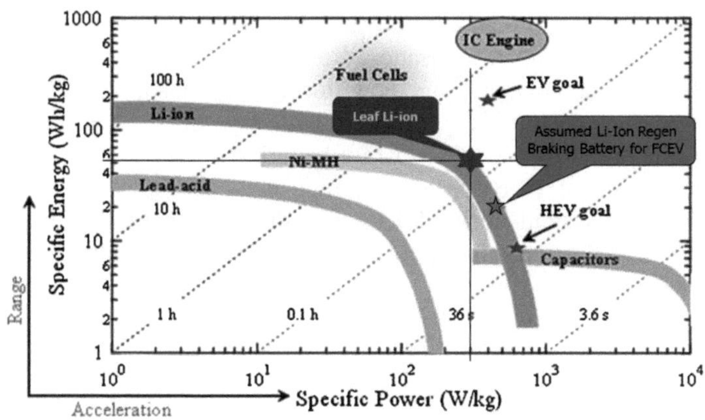

Fig. A.1 Ragone chart for various batteries, showing the relation between the Nissan Leaf high-energy battery and the high-power battery for a FCEV

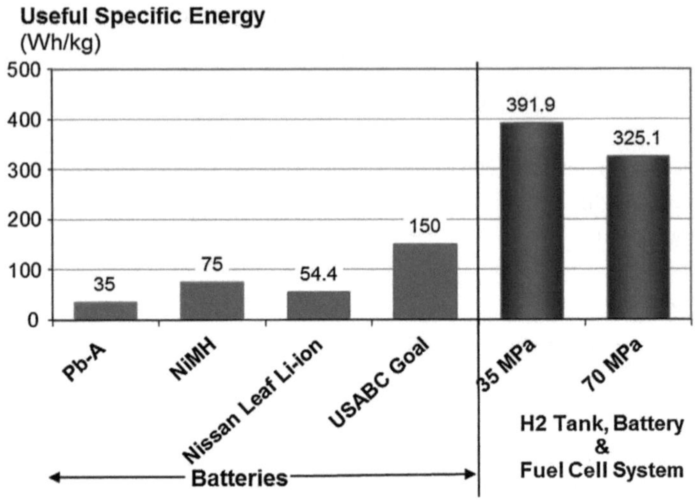

Fig. A.2 Useful specific energies for battery and fuel cell systems

structure, brakes, and suspension systems. This added mass will require still more batteries to provide the desired range.

The mass for any electric vehicle is limited by the useful specific energy (in Wh kg^{-1}) of the complete storage system. As shown in Fig. A.2, the specific energy of battery systems has improved over the last few decades, from 35 Wh kg^{-1} for the lead-acid (Pb-A) batteries used to start ICVs for a century to 75 Wh kg^{-1} for nickel-metal hydride (NiMH) batteries used in the original Prius HEVs, to the advanced lithium ion (Li-ion) batteries used in laptop computers, cell phones, and now BEVs such as the Nissan Leaf BEV; the USABC goal for BEV batteries is 150 Wh kg^{-1}. As shown in Fig. A.2, the specific energy for a fuel cell energy storage system (hydrogen tanks, plus the fuel cell system plus a peak power battery system) is better (larger) than even advanced Li-ion batteries that meet the minimum goals for the US Advanced Battery Consortium long-term commercialization goals [A.3]. Most car companies are now storing hydrogen at 70 MPa (700 bar) on their FCEVs. The specific energy of these hydrogen storage systems is 1.3 times the specific energy of the USABC goal for batteries.

The breakdown of FCEV energy storage mass is shown in Fig. A.3 for a 70-MPa (700-bar) system storing 5.1 kg of hydrogen.[4] The hydrogen tank storage system mass is 133 kg, accounting for 48 % of the FC system mass.

[4]This amount of hydrogen (5.1 kg) is sufficient to provide a range of approximately 350 miles for this FCEV. The hydrogen plus tank mass would be much less for the shorter ranges associated with BEVs.

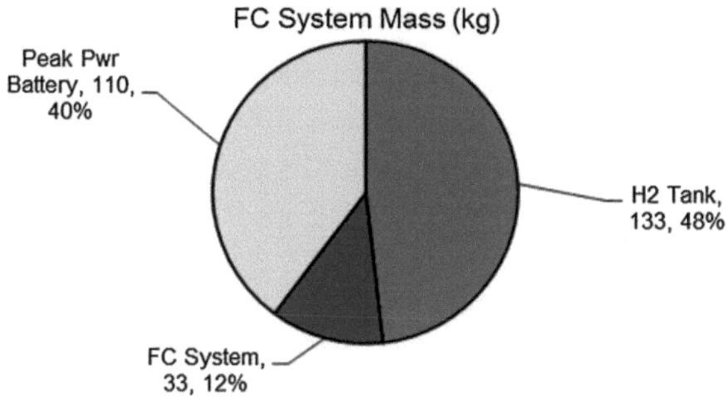

Fig. A.3 Breakdown of FC storage system component masses

Mass Compounding for BEVs

According to Malen and Reddy [A.2], the final mass of a vehicle after mass compounding is given by

$$M_\mathrm{f} = M_\mathrm{i} + \Delta + \Delta G \tag{A.1}$$

where

M_i is the initial vehicle mass,
M_f is the final vehicle mass,
Δ is the change in mass (e.g., due to increased battery or motor/controller mass), and
G is the secondary vehicle mass coefficient

$$G = \gamma/(1-\gamma) \tag{A.2}$$

where $\gamma =$ the sum of the influence coefficients for all vehicle components which Malen and Reddy measured as 0.39 for the average of late model vehicles. In this case,

$$G = \frac{0.39}{(1-0.39)} = 0.6393 \tag{A.3}$$

For the initial vehicle mass, we use an aluminum-intensive vehicle (AIV Sable) developed by the Ford Motor Company in the mid-1990s as part of the partnership for a new generation of vehicles (PNGVs) between the auto companies and the Clinton administration to improve fuel economy by lowering the mass of vehicles.[5] The estimated mass of this vehicle without the drivetrain (the "glider") is estimated at 825 kg as summarized in Table A.2.

To determine the mass of a BEV based on this lightweight glider, we need to add a battery bank and a motor/controller system and associated electrical

[5]Ford developed their first FCEVs under this same, PNGV program.

Table A.2 Estimated mass of a lightweight BEV glider

AIV Sable design basis	Mass (kg)
Test weight	1291
ICE	195
Fuel system	64.9
Exhaust, cat converter, and muffler	30.8
Transmission and Differential	95.7
Driver	80
Estimated glider mass	824.6

equipment. The size of the battery is determined by the total mass of the vehicle and the design range target for the vehicle. We have previously developed a vehicle drive simulation model for electric vehicles both battery- and fuel cell-powered under contract to the Ford Motor Company. The total amount of energy required per kilometer traveled for the motor/controller is shown in Fig. A.4 as a function of the vehicle test mass. The upper dashed line is the energy per mile required with a 1-kW AC load plus a 500-W accessory load, and the lower solid line assumes no AC but with a 500-W accessory load. For comparison, the triangle shows the measured energy per km for the Nissan Leaf BEV based on a curb mass of 1521 kg plus 80 kg for the driver or a total of 1601 kg test mass, and the square shows the data for a 2012 Tesla Model S with their large 85-kWh battery and a test mass of 2163 kg as tested by Edmunds, showing an excellent match between experimental data and our simulation model. Edmunds, an independent testing company, measured an energy requirement of 0.213 kWh/km (0.343 kWh/mile) for the Leaf

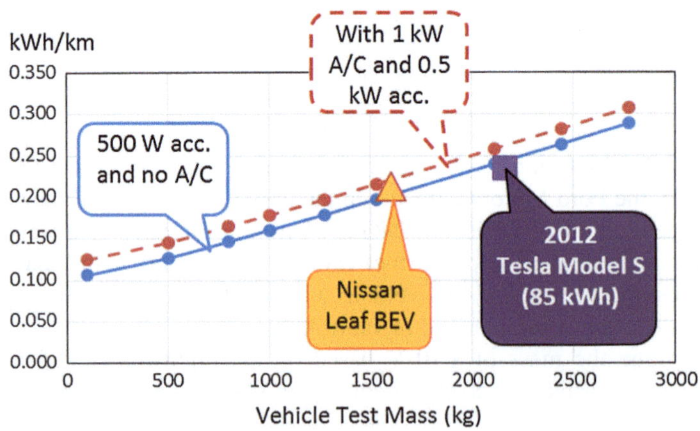

Fig. A.4 Energy per km required to the motor/controller of an electric vehicle as a function of vehicle test mass

BEV [A.4]. The required motor/controller input energy per mile is given by a linear fit to the upper line in Fig. A.4:

$$E_b = m_1 \times M_f + b_1 \tag{A.4}$$

where

E_b is the energy per km required from the battery pack (kWh/km)
M_f the final test mass of the vehicle
m_1 the slope of the energy/kg versus Mf line = 6.959×10^{-5} kWh/kg-km, and
b_1 the intercept of the linear curve fit. = 0.111667 kWh/km

The mass of the battery pack will increase for longer ranges; for the initial mass, we assume a range of 117.5 km (73 miles—the EPA-rated range of the Leaf BEV).

The resulting initial mass of the battery for a range of 117.5 km is then

$$M_b = E_b \times R/\mathrm{SE} \tag{A.5}$$

where SE is the specific energy of the battery pack = 0.0544 kWh/kg from Table A.1. The initial mass of the BEV at 117.5 km design range is given by

$$M_i = M_b + M_{mc} + M_g + M_L \tag{A.6}$$

where

M_{mc} is the mass of the motor/controller,
M_g the glider mass = 824.6 kg from Table A.2, and
M_L the vehicle load (passengers and cargo)

The vehicle load term is needed to account for the mass of passengers and cargo. At the very least, there will always be one "passenger": the driver! The size of the battery bank and motor system will need to be increased to accommodate this extra mass. The power required from the vehicle motor is set by the power

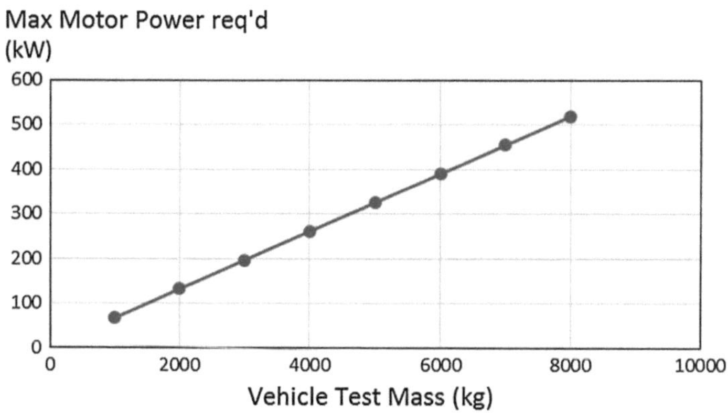

Fig. A.5 Motor power required for 10-s acceleration from zero to 96.5 km/h (60 mph) as a function of vehicle test mass

required to accelerate from zero to 96.5 km/h (60 mph) in 10 s,[6] as shown in Fig. A.5. The required motor power is given by a linear fit to those data:

$$P_{\mathrm{m}} = m_2 \times M_{\mathrm{f}} + b_2 \tag{A.7}$$

where

Pm is the required peak power (kW)
m_2 is the slope of the motor power curve = 0.06459 kW/kg
b_2 the y-intercept = 3.12 kW

Given the required motor power, we can estimate the mass of the motor and controller from various motor manufacturers as summarized in Fig. A.6. In this case, there is much scatter in the data, so we take the upper bound shown in Fig. A.6 for motors with greater than 40 kW of power (which covers all BEVs considered here).

The mass of the motor/controller is then given approximately by:

$$M_{\mathrm{mc}} \approx m_3 \times P_{\mathrm{m}} + b_3 \tag{A.8}$$

where

m_3 is the slope of the topline = 0.04357 kg/kW
b_3 is the intercept of that line = 92 kg

Substituting the above equations in Eq. A.6, we can solve the initial BEV test mass:

$$M_{\mathrm{i}} = (m_1 M_{\mathrm{i}} + b_1) \times R/\mathrm{SE} + m_3(m_2 M_{\mathrm{i}} + b_2) + b_3 + M_{\mathrm{g}} + M_{\mathrm{L}} \tag{A.9}$$

or

$$M_{\mathrm{i}} = \left[M_{\mathrm{g}} + M_{\mathrm{L}} + b_1 R/\mathrm{SE} + m_3 b_2 + b_3 + M_{\mathrm{g}}\right]/[1 - m_1 R/\mathrm{SE} - m_2 m_3] \tag{A.10}$$

The parameters for calculating the initial BEV mass via Eq. A.10 are summarized in Table A.3 where we assume only one passenger—the driver with a mass of 80 kg. For a range of 117.5 km,[7] the initial battery would have a mass of 460.7 kg with the Nissan Leaf specific energy of 54.4 Wh/kg, and the total BEV test mass would be 1462 kg.

Now that we have calculated the initial test mass of the BEV based on the Nissan Leaf, we can resume using Eq. A.1 to calculate how the BEV mass increases with increasing range above the 117.5 km assumed design range.

The "Δ" in Eq. A.1 refers to the increased mass due to larger batteries and larger motors. The battery mass (Eq. A.5) increases linearly with vehicle mass, and the motor/controller mass increases linearly with motor power (Eq. A.8), which in turn increases linearly with range (A.7).

---

[6]The vehicle motor must also have enough power for actions such as hill-climbing and accelerating at lower speeds, etc., but the zero to 37 km/s acceleration in 10 s requires the highest power.

[7]The EPA lists the range of the Nissan Leaf at 73 miles or 117.5 km.

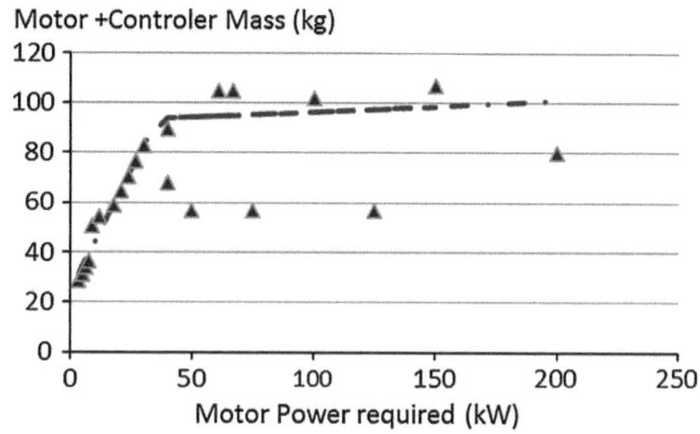

Fig. A.6 Motor/controller mass versus power required

Table A.3 Parameters for calculating the initial mass of the Nisan Leaf BEV with 117.5 km range

	#1	#2	#3
	kWh/km	M_{tr} P_{wr} (kW)	M_{mc} (kg)
m	6.959E-05	0.06460	0.04357
b	0.1117	3.1209	92.0080
R	117.457	km	
SE	0.0544	kWh/kg	
M_g	824.6	kg	
M_b	460.7	kg	
P_m	97.5	kW	
M_t	0.00	kg	
M_{fc}	0.00	kg	
M_{mc}	96.3	kg	
M_i	1461.5	kg	

Therefore, the BEV change in mass is given by the difference between the component masses at a range "R" and the initial value for that component at $R = 117.5$ km:

$$\Delta = M_b - 460.7 + M_{mc} - 96.3 \text{ kg}. \tag{A.11}$$

Substituting these equations into Eq. A.1 yields the final BEV mass as a function of range including the effects of mass compounding:

$$M_f = M_i + \Delta C = M_i + (M_b + M_{mc})C - (460.8 + 96.7)C = (M_b + M_{mc})C - 556.9C \tag{A.12}$$

$$= M_i + [m_1 M_f + b_1)RC/\text{SE} + m_3(m_2 M_f + b_2) + b_3)]C - 556.9C \tag{A.13}$$

Or

$$M_f = M_i + M_L + [b_1 R/\text{SE} + m_3 b_2 + b_3 - 556.9]C/[1 - m_1 RC/\text{SE} - m_2 m_3 C] \quad \text{for BEVs} \tag{A.14}$$

where $C = 1 + G = 1.6393$.

Mass Compounding for FCEVs

FCEVs are also subject to mass compounding, although the effects are quite small compared to BEVs, since adding extra range for a FCEV means enlarging the hydrogen tanks, which weigh much less than batteries per kWh of stored energy. Almost all car companies now store hydrogen at high pressure in carbon fiber-wrapped tanks, the same technology that has been used to store natural gas in natural gas vehicles for more than 50 years. Hydrogen is typically stored at approximately 70 MPa ($\approx$10,150 psi) compared to 24.8 MPa (3600 psi) for NGVs.

The weight percent of hydrogen in carbon fiber-wrapped tanks is shown in Fig. A.7 for two vendors of 700-bar tanks. The weight percent of hydrogen is in the range of 4 % or higher for tanks that store 1–5 kg of hydrogen. Five kg of hydrogen is enough to travel up to 212 km (342 miles) given the measured fuel economy of 42.4 km/kg (68.3 miles/kg) measured by the DOE National Laboratories for the Toyota Highlander SUV FCEV [A.6]. Hydrogen content will be less for shorter range vehicles in the calculations below, and FCEV fuel economy should rise with smaller vehicles and more recent fuel cells that should have higher efficiency and higher fuel economy.

The weight percents shown in Fig. A.7 do not take into account the hydrogen remaining in the tank. A vacuum pump would have to be installed to pull all the hydrogen out of the tank, which would not be cost-effective. Therefore some hydrogen will necessarily remain in the tank when the driver pulls into the station to fill up more hydrogen (this is analogous to not totally depleting the battery bank on a BEV.)[8] Figure A.8 shows the impact of leaving some hydrogen in the tank. The solid line shows the useful hydrogen from a tank with a water volume of 29.2 l as a function of the minimum tank pressure left in the tank. In this particular case, the vendor lists a hydrogen capacity of 1.2 kg (upper dashed line), but this water volume would only hold 1.15 kg even if all hydrogen were removed. With a minimum residual pressure of 2 bar (29 psig), this particular tank could deliver only 1.14 kg of useful hydrogen to the fuel cell.

The net result is that a tank designed to hold 4 % hydrogen by weight will yield slightly less hydrogen taking into account the nonzero residual pressure for an "empty" tank as shown in Fig. A.9. We assume an "empty" tank pressure of 20 bar (29 psig) which yields a net useful hydrogen weight percent of 3.84 % for a tank with a nominal 4 % by weight hydrogen capacity.

The volume occupied by these tanks is limited by the compressibility of hydrogen. If hydrogen were an ideal gas, then the amount of hydrogen in a tank would double if the pressure was increased from 350 bar to 700 bar. However, hydrogen is not an ideal gas, as shown in Fig. A.10. Instead of doubling the internal mass of stored hydrogen by doubling the pressure from 350 to 700 bar, the internal mass of stored hydrogen increases by only 67 %.

[8] With the major exception that a depleted battery either significantly shortens the battery life, or even destroys the battery, a depleted hydrogen tank suffers no degradation.

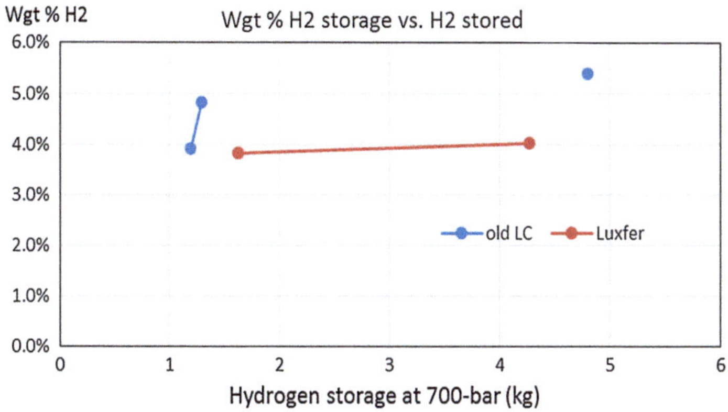

Fig. A.7 Weight percent of hydrogen versus kg of 700-bar hydrogen stored ("Old LC" refers to the Lincoln Composites which is now Hexagon Lincoln; Luxfer is the composite tank fabricator that acquired Dynatek, a former leading Canadian composite tank manufacturer)

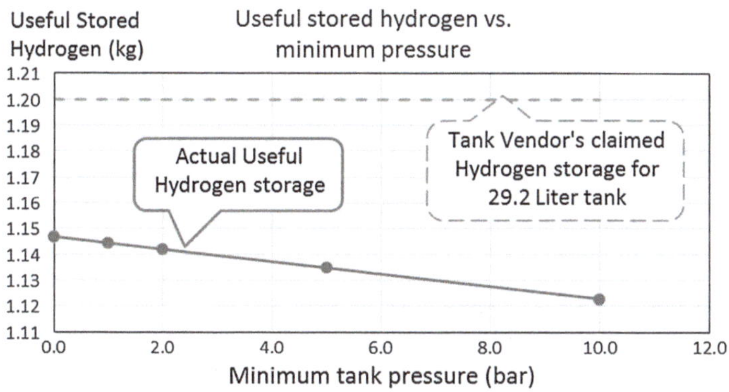

Fig. A.8 Useful stored hydrogen for a 700-bar compressed hydrogen tank with a water volume of 29.2 l as a function of the residual pressure in the tank when it is refilled

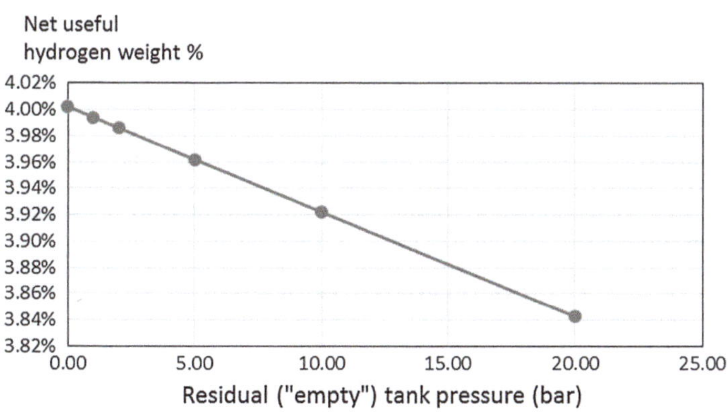

Fig. A.9 Net useful hydrogen weight % as a function of residual empty tank pressure

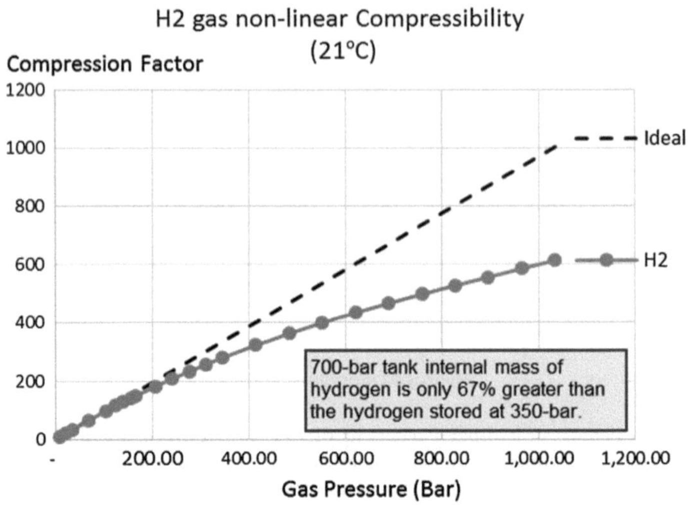

Fig. A.10 Nonlinear compressibility of hydrogen compared to an ideal gas

With the assumed 4 % hydrogen weight percent for an *empty* 700-bar tank and a 3.84 % useful weight percent assuming a 20-bar (29 psig) residual pressure in an "empty" tank, then the mass of the tank plus the hydrogen is given by:

$$M_t = (1 - 3.84\ \%)/3.84\ \% \times M_h = 25.022 \times M_h \tag{A.15}$$

where

M_t is the mass of the hydrogen plus the tank and
M_h is the mass of the hydrogen

The amount of hydrogen required for a given range and vehicle test mass is derived from the vehicle simulation model by calculating the kg of hydrogen required per km traveled in the FCEV, as shown in Fig. A.11. The hydrogen required per km is given by:

$$M_h = (m_4 M_f + b_4) \times R(\text{km}) \tag{A.16}$$

where

m_4 slope of line $= 3.79 \times 10^{-6}$ in Fig. A.7
M_f the final vehicle test mass (kg), and $b_4 = 0.00539$ kg of H2/km

Combining Eqs. A.15 and A.16 yields the combined mass of the tank plus hydrogen as a function of the vehicle test mass and range:

$$M_t = (m_5 M_f + b_5) \times R(\text{km}) \tag{A.17}$$

where

$m_5 = 9.48 \times 10^{-5}\text{kg}^{-1}\text{km}^{-1}$
$B_5 = 0.13487$ kg/km

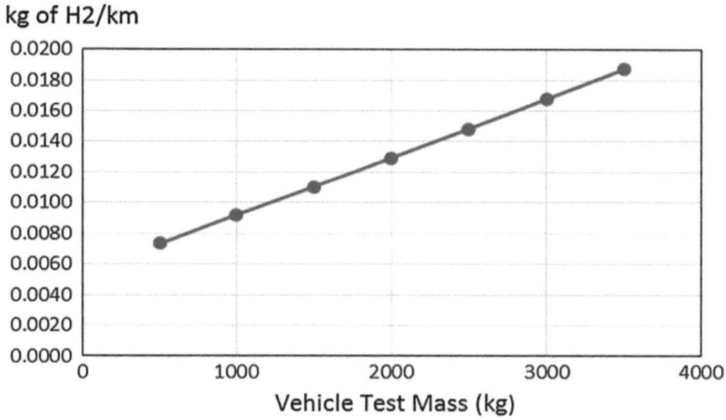

Fig. A.11 Kilogram of hydrogen required per km of range versus vehicle test mass

The power required from the fuel cell system and the peak power battery can also be represented by linear fits to the vehicle drive cycle simulations given by Eq. A.7; this peak power for acceleration can be provided by both the fuel cell system and the peak power battery. We assume here that 90 % of the power for normal driving is provided by the fuel cell and 10 % by the peak power battery. The total power from the fuel cell from Eq. A.7 is then given by:

$$P_{fc} = 0.9 \times (m_2 M_f + b_2) = \text{power required from the fuel cell system} \tag{A.18}$$

m_2 is the slope of the motor power curve = 0.06459 kW/kg,

and

b_2 the y-intercept = 3.12 kW,

and

$$P_b = 0.1 \times (m_2 M_f + b_2) = \text{power required from the peak power battery} \tag{A.19}$$

The mass of the peak power battery becomes:

$$M_b = P_b / SP_b \tag{A.20}$$

where SP_b = the specific power of the peak power battery (0.4 kW/kg).

The mass of the fuel cell system is determined by the fuel cell specific power in kW/kg. The specific power of fuel cell systems has been increasing over time as companies improve their technology. In the late 1990s, fuel cell specific power was in the one kW/kg range, but Nissan announced a new fuel cell system in October 2011 that delivered 85 kW with a mass of only 43 kg or a specific power of 1.98 kW/kg [A.6] and a volume of only 34 l or a power density of 2.5 kW/l, which we use here, with the expectation that improvements will reduce the weight and volume of future fuel cell systems. The resulting mass of the fuel cell system becomes:

$$M_{fc} = P_{fc}/SP_{fc} \tag{A.21}$$

where SP_{fc} = the specific mass of the fuel cell system (1.98 kW/kg)

To calculate the initial mass of the FCEV, we use an equation similar to Eq. (A.6) that we used to calculate the initial BEV mass above. We assume that the FCEV uses the same glider as the BEV. The initial FCEV mass becomes:

$$M_i = M_g + M_{fc} + M_b + M_t + M_{mc} + M_L \tag{A.22}$$

where M_L = the vehicle load (at least 80 kg for the driver.)

Or

$$M_i = M_g + M_L + 0.9/SP_{fc} \times (m_2 M_i + b_2) + (1 - 0.9)/SP_b \times (m_2 M_i + b_2) + (m_5 M_i + b_5) \times R + m_3(m_2 M_i + b_2) + b_3 \tag{A.23}$$

To simplify these equations, define a new constant, Q:

$$Q \equiv 0.9/SP_{fc} + 0.1/SP_b \tag{A.24}$$

Which leads to the following equation for the initial FCEV mass at the same range as the BEV (117 km):

$$M_i = [M_g + M_L + Qb_2 + b_5 R + b_3 + m_3 b_2]/(1 - Qm_2 - m_5 R - m_3 m_2) \tag{A.25}$$

With the parameters listed above, the initial mass of the FCEV for 117 km range is 1079 kg, which is lighter than any of the other AFVs at that range as shown in Table A.4.

Next we need to solve for the change in mass, Δ, to solve Eq. A.12 for the final FCEV mass as a function of range. The four main FCEV components initially weigh 174.4 kg as summarized in Table A.5.

To solve Eq. A.12 for the final FCEV mass requires a formula for the change in mass, Δ:

Table A.4 Test mass of the alternative vehicles at a range of 117 km

	Mass (kg)
FCEV	1079
USABC goal BEV	1150
NiMH BEV	1320
Leaf Li-ion BEV	1462
Pb-A BEV	1796

Table A.5 Initial mass (kg) of FCEV drivetrain components

Fuel cell system	33.2
Peak power battery	18.2
Motor controller	95.2
Hydrogen + tank	27.9
Total drivetrain mass	174.4

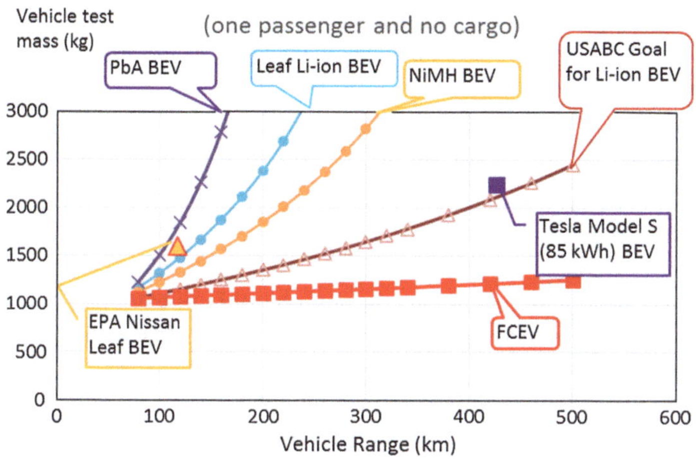

Fig. A.12 BEV and FCEV mass versus range assuming one driver and no cargo

$$\begin{aligned}\Delta &= M_{\mathrm{fc}} + M_{\mathrm{b}} + M_{\mathrm{mc}} + M_{\mathrm{t}} - 174.4 \\ &= \mathrm{QP_m} + m_3 P_{\mathrm{m}} + b_3 + (m_5 M_{\mathrm{f}} + b_5) R(\mathrm{km}) - 174.4 \\ &= (Q + m_3) P_{\mathrm{m}} + b_3 + (m_5 M_{\mathrm{f}} + b_5) R - 174.4\end{aligned} \tag{A.26}$$

From Eq. A.12, we have:

$$\begin{aligned}M_{\mathrm{f}} &= M_{\mathrm{i}} + \Delta C \\ &= M_{\mathrm{i}} + C[(Q + m_3) \times (m_2 M_{\mathrm{f}} + b_2) + b_3 + (m_5 M_{\mathrm{f}} + b_5) R - 174.4]\end{aligned} \tag{A.27}$$

Or

$$M_{\mathrm{f}} = [M_i + [(Q + m_3) b_2 + b_3 + b_5 R - 174.4] \times C]/[1 - (Q + m_3) m_2 C - m_5 R C] \tag{A.28}$$

Equation A.14 for BEVs and Eq. A.27 for FCEVs are plotted in Fig. A.12. The FCEV mass changes very little with range, while the mass of the BEVs grows rapidly with range as the battery bank mass increases. Even if batteries are developed in the future that meet the USABC commercialization goals, the BEV will still have a much greater mass than a FCEV. Two datum points are shown in Fig. A.12 representing the curb mass of the Nissan Leaf plus driver (763 kg (1601 lbs) and the EPA estimated range of 117 km (73 miles). This point is above our estimated mass for a BEV with Leaf battery parameters, indicating that our equations are slightly conservative (projecting lower mass than this one example.)

The second BEV point is for the Tesla Model S (85-kWh battery version), which Edmunds tested with a curb weight of 2163 kg (4770 lbs) [A.7] to which we add 80 kg for the driver and which the EPA has certified a range of 426 km (265 miles). The Tesla point falls just above the USABC goal line for advanced Li-ion batteries. However, the EPA data on the Tesla Model S are not valid [A.8].

Table A.6 Battery parameters used in Fig. A.12

Battery type	Specific energy (kWh/kg)	Energy density (kWh/l)
Pb-A BEV	0.035	0.095
NiMH BEV	0.075	0.135
Leaf Li-ion BEV	0.054	0.324
USABC goal BEV	0.15	0.324

For example, they claim an energy consumption of 0.38 kWh/mile and a range of 265 miles. But this would require $0.38 \times 265 = 100.7$ kWh of battery energy, but the Tesla Model S battery reportedly holds only 85kWh of energy. Even if the 85 kWh battery were completely discharged, the range would be only $85/0.38 = 224$ miles, which is still very good for a BEV. However, the Edmunds testing company did achieve a 267 mile range on one Model S test drive with a nominal 85-kWh battery pack [A.7], so either the Tesla battery holds more than 85 kWh or the fuel economy is better (lower) than 0.38 kWh/mile specified by the EPA.

The battery parameters used in Fig. A.12 are summarized in Table A.6.[9]

Energy Storage Volume

Most readers know that batteries are heavy, but some may not realize that batteries also take up substantial space. We have also calculated the volume required for energy storage for BEVs and FCEVs as summarized in Fig. A.13. In the case of BEVs, the storage volume is just the space occupied by the battery bank. For FCEVs, the storage volume is taken as the sum of the volumes for the hydrogen storage tank, the peak power battery, and the fuel cell system itself: In other words, all the components that are required to deliver electricity to the motor/controller as a replacement for the battery bank.

The battery energy densities are taken from Table A.6. The FCEV component volumes are estimated as follows:

- FC system: based on Nissan FC with 85 kW and 34 l or 2.5 kW/l [A.5]
- Peak power battery: high-power battery (400 W/kg) but low energy (22 Wh/kg); we assume a Li-ion energy density of 324 Wh/l or 0.0679 l/kg of battery.
- Hydrogen fuel tank: The fuel tank volume is given by:

$$V_t = 30.7M_h + 3.23 \quad (A.29)$$

where M_h = the quantity of hydrogen required (kg) (see Eq. A.16)

[9]Note: the energy density goal for the USABC commercialization goal is only 0.230 kWh/l; since the Nissan Leaf battery has already demonstrated 0.324 kWh/l, we increased the energy density goal to that value.

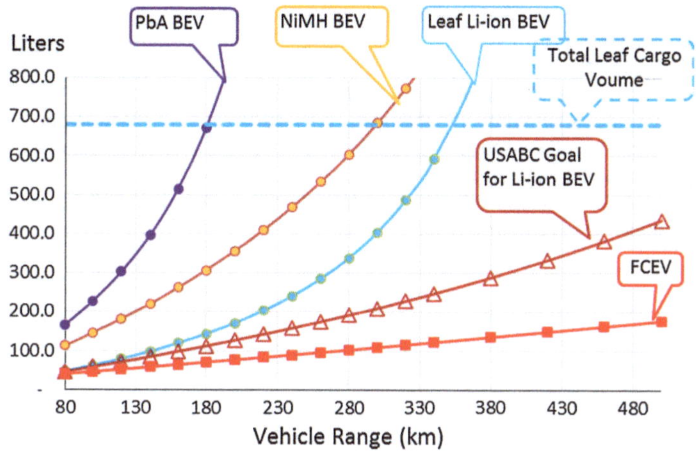

Fig. A.13 Onboard storage volume versus range for BEVs and FCEVs

As shown in Fig. A.13, the required volume for storing batteries on BEVs grows very rapidly as range increases. The horizontal dashed line at the top of the figure corresponds to the total cargo storage available on the Nissan Leaf BEV, listed as 679 l (24 ft^3). Thus, if one tried to build a BEV with Pb-A batteries, then the entire cargo space on the Leaf would be occupied above a range of 180 km (112 miles), while a BEV with NiMH batteries would run out of cargo space at 300 km (186 mile) range, and the Leaf itself would fill all its cargo space to achieve a range of approximately 350 km (217 miles).

If a battery technology could achieve the USABC commercialization goals, then ranges above 500 km (311 miles) would be feasible, although, even then, the volume required for the entire hydrogen and fuel cell system would still occupy only 41 % of the space required for an advanced battery system meeting USABC goals with a 500 km range.

References

[A.1] Nissan Leaf spec-sheet FINAL us 2.pdf, available at http://www.electricvehiclewiki.com/File:NISSAN_LEAF_SpecSheet_FINAL_US_2.pdf

[A.2] D. E. Malen & K. Reddy, "Preliminary vehicle mass estimation using empirical subsystem influence coefficients," University of Michigan, May 9, 2007 (revised June 26, 2007), availableat:http://www.a-sp.org/database/custom/Mass20Compounding20-20Final20Report.pdf.

[A.3] See USABC Goals for Advanced Batteries for EVs on the U.S. Council for Automotive Research, LLC web page at: http://www.uscar.org/guest/article_view.php?articles_id=85.

[A.4] Edmunds, Inc. "2011 Nissan Leaf Road Test 2" available at: http://www.edmunds.com/nissan/leaf/2011/road-test3.html

[A.5] Nissan fuel cell stack achieves world-beating power density," available at http://integrityexports.com/2011/10/14/nissan-fuel-cell-stack-gets-world-beating-power-density/
[A.6] K. Wipke, D. Anton, & .S. Sprik, "Evaluation of range estimates for Toyota FCHV-adv under open-road conditions," Report # SRNS-STI-2009-00446, August 2009, available at: http://www.nrel.gov/hydrogen/pdfs/toyota_fchv-adv_range_verification.pdf
[A.7] D. Edmunds, "An EV fit for a Sunday drive," 9/19/2012, available at: http://www.edmunds.com/tesla/model-s/2012/road-test1.html
[A.8] Side-by-side comparisons, Environmental Protection Agency available at http://www.fueleconomy.gov/feg/evsbs.shtml

Appendix B
Distribution of Vehicles by Class

Appendix A demonstrated that BEVs will be limited to small cars traveling short distances, since both the weight and volume of batteries for larger cars traveling longer distances are excessive. This appendix explores the consequences of this limitation in terms of the number of light-duty vehicles (LDVs) affected and their contribution to greenhouse gas (GHG) emissions and oil consumption in the USA.

The fraction of small cars and trucks sold in the USA has been decreasing as shown in Fig. B.1, dropping from ≈40 % of market share in the 1980s to only 28.8 % in 2012 as American drivers began buying larger cars, wagons, SUVs, vans, and pickup trucks. The fraction of GHG emissions generated by smaller cars and trucks has also been decreasing from approximately 40 % of all new car and truck emissions in the 1980s to 26.3 % in 2012 [B.1]. So to the degree that the distribution of new cars sold approximates the distribution of cars on the road (OTR), replacing small cars with BEVs would affect only 26 % of greenhouse gases.

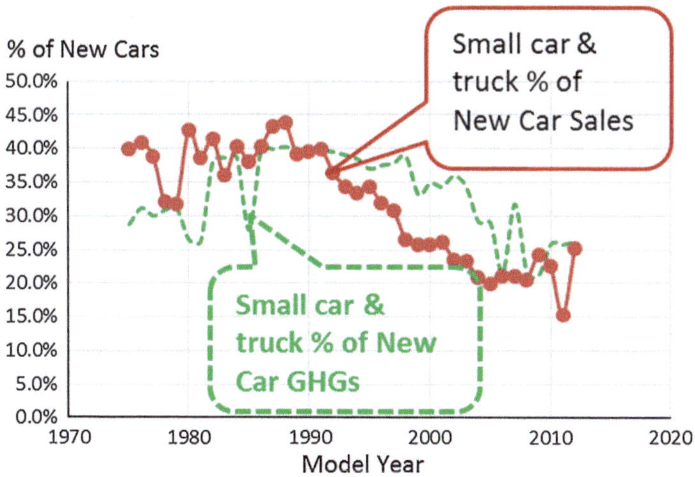

Fig. B.1 Fraction of new small cars and trucks sold each year and the fraction of GHG emissions from those small cars and trucks compared to all new cars sold each year

C.E. (Sandy) Thomas, *Sustainable Transportation Options for the 21st Century and Beyond*, DOI 10.1007/978-3-319-16832-6

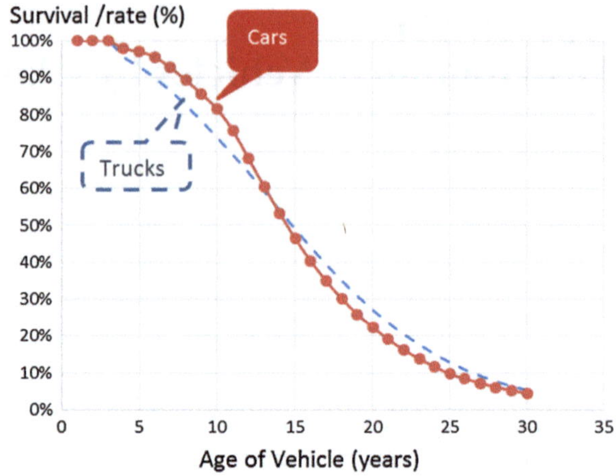

Fig. B.2 Survival rates for light-duty cars and trucks

The emissions listed in Fig. B.1 are the fraction of GHGs attributed to *new cars* compared to all new cars sold that year. It does not indicate the GHG emissions from all cars on the road (*OTR*). To estimate the fraction of GHGs attributable to small cars and trucks OTR, we need to determine the mixture of vehicles OTR over time.

The Oak Ridge National Laboratory has published lists of survival rates [B.2] for light-duty cars[10] and trucks[11] as shown in Fig. B.2. New cars and trucks survive for several years. On the average, only 4.5 % of cars survive after 30 years, and only 5.5 % of light-duty trucks are OTR after 30 years. By multiplying the annual new car and truck sales times the appropriate survival rates, we can estimate the number of LDVs OTR by vehicle class. As shown in Table B.1, the estimated percentage of all small cars and trucks[12] OTR was only 37.9 % in 2012, and the percentage of all small cars and trucks sold in 2012 was only 28.8 %.[13]

In addition, not all "small" cars are really suitable for becoming a battery EV. For example, the General Motors' Cadillac GTS Wagon is classified as a "small wagon" by the EPA which is based solely on its passenger volume of 3160 l (111.6 ft^3).[14] But the Cadillac GTS is a very heavy car with a curb weight of 1826 kg (4025 lbs), which would not be a good candidate for a battery EV without a major reduction in curb weight. Assuming a test weight of 1904 kg,[15] this vehicle would initially require a battery delivered energy of 0.226 kWh/km (0.364 kWh/mile) (see Appendix A). Assuming a range of 161 km (100 miles), the battery bank

---

[10]Light-duty cars include sedans, wagons, and car-based SUVs.

[11]Light-duty trucks include pickup trucks, vans, and truck-based SUVs.

[12]Small vehicles include all small pickup trucks, small vans, small wagons and small SUVs.

[13]*The last row of* Fig. B.2 is the average of just the small vehicles.

[14]EPA defines a small wagon as having less than 3681 l (130 cu. Ft.) of passenger and cargo space.

[15]Curb weight plus 80 kg for the driver.

Table B.1 Percentage of vehicles on the road and vehicles sold in 2012 according to size class (T = Truck based & NT = non-truck based)

	Vehicles on the road (%)	2012 % sales	Average OTR and sales (%)
Small car	24.7	25.1	24.9
Midsize car	18.1	21.7	19.9
Large car	8.3	6.2	7.3
Small wagon	2.2	3.6	2.9
Midsize wagon	1.0	0.3	0.7
Large wagon	0.2	0.0	0.1
Small NT SUV	0.3	0.0	0.1
Midsize NT SUV	1.5	4.0	2.7
Large NT SUV	0.8	2.9	1.8
Small pickups	2.9	0.0	1.46
Midsize pickup	10.6	0.2	5.4
Large pickup	0.1	9.8	4.9
Small van	6.7	0.0	3.35
Midsize van	0.6	4.9	2.8
Large van	10.3	0.2	5.2
Small T SUV	1.1	0.0	0.5
Midsize T SUV	9.0	10.3	9.7
Large T SUV	7.4	10.6	9.0
All small vehicles	37.9	28.8	33.3

would have to initially deliver 36.4 kWh of energy to travel 161 km on one charge. Assuming that battery manufacturers achieve the USABC specific energy goal of 150 Wh/kg, then the battery bank would have an initial mass of 36.4/.15 = 243 kg. After mass compounding, the final Cadillac BEV vehicle test mass would become[16]:

$$M_{\mathrm{f}} = 1909 + 242 \times 1.6393 = 2306 \text{ kg}.$$

This higher BEV test mass after mass compounding would then require a larger battery bank that could deliver 0.254 kWh/km (0.408 kWh/mile) or a final battery energy capacity of 40.8 kWh for a 161-km (100 mile) range. From Chap. 6, the battery alone would then cost between $6120 assuming that the battery industry meets the USABC battery cost goal of $150/kWh and as much as $20,400 for the battery alone by 2020 with other estimates of lithium-ion battery costs in the range between $200/kWh and $500/kWh by the 2020 time frame. These high battery costs could inhibit the purchase of heavy BEVs such as the Cadillac GTS Wagon and might also preclude affordable battery warranties that would also limit BEV sales.

[16]See Eq. A.1 in Appendix A.

Table B.2 Estimated percentage of potential BEVs

	Vehicles on the road (%)	2012 % sales	Average OTR and sales (%)	% that could be BEVs	% that have access to off-street parking	Net BEV percentages
Small car	24.7	25.1	24.9	100	56	14.0
Midsize car	18.1	21.7	19.9	50	56	5.6
Large car	8.3	6.2	7.3	30	56	1.2
Small wagon	2.2	3.6	2.9	95	56	1.6
Midsize wagon	1.0	0.3	0.7	50	56	0.2
Large wagon	0.2	0.0	0.1	30	56	0.0
Small NT SUV	0.3	0.0	0.1	90	56	0.1
Midsize NT SUV	1.5	4.0	2.7	40	56	0.6
Large NT SUV	0.8	2.9	1.8	25	56	0.3
Small pickups	2.9	0.0	1.46	80	56	0.7
Midsize pickup	10.6	0.2	5.4	30	56	0.9
Large pickup	0.1	9.8	4.9	20	56	0.6
Small van	6.7	0.0	3.35	80	56	1.5
Midsize van	0.6	4.9	2.8	30	56	0.5
Large van	10.3	0.2	5.2	20	56	0.6
Small T SUV	1.1	0.0	0.5	80	56	0.2
Midsize T SUV	9.0	10.3	9.7	30	56	1.6
Large T SUV	7.4	10.6	9.0	20	56	1.0
All small vehicles	37.9	28.8	33.3	Total estimated BEV potential:		31.0

We conclude that not all "small wagons" could be affordably converted to run on batteries.

Access to overnight parking spaces. In addition to high mass limitations, BEV sales will be hindered by the lack of a parking space to park the vehicle overnight. One main benefit of a BEV is the opportunity to "refuel" by plugging in the BEV overnight at the driver's residence. Researchers at Carnegie Mellon University have estimated that only 56 % of US motor vehicles have access to a dedicated parking space suitable for overnight charging and that only 22 % of all vehicles have access to a charging outlet [B.3]. Therefore, even if they have a dedicated parking space, they would still have to install an appropriate electrical outlet near that parking space.

The total percentage of US vehicles that could be converted to BEVs is estimated in Table B.2 by multiplying the percentage of vehicles OTR and the percentage that could be affordably converted to BEVs and the percentage of those that have access to off-street parking for nighttime charging, with the final estimate that approximately 31 % of all US motor vehicles (light-duty cars and trucks) could become BEVs in the future.

References

[B.1] S. C. Davis, S. W. Diegel & R. G. Boundy, "Transportation Energy Data Book: Edition 32," Oak Ridge National Laboratory, Table 4.11 (new car & truck sales) and Table 11.8 (GHG emissions.) available at: http://cta.ornl.gov/data/index.shtml

[B.2] Op. cit., car survival rates, Table 3.12; truck survival rates, Table 3.13.

[B.3] E.J. Traut et al., "U.S. residential charging potential for electric vehicles," Carnegie Mellon University, *Transportation Research Part D* 25 (2013) 139-145. Available at: http://www.cmu.edu/me/ddl/publications/2013-TRD-Traut-etal-Residential-EV-Charging.pdf

Appendix C
Estimated Future US Light-Duty Vehicle Sales, Vehicles on the Road, and Number of Miles Driven Annually

The impact of adding alternative vehicles on the environment and oil consumption will depend on the total number of vehicles on the road in the future and the efficiency and number of miles that those vehicles are driven. Each of these factors grew rapidly in the second half of the twentieth-century with growth rates that cannot be sustained. We have to select suitable growth rates over the twenty-first century to model the impact of alternative vehicles. These estimates are not meant to precisely predict the future, but rather to postulate a set of *plausible* vehicle use parameters over the century as a basis for judging the potential impact of adding different alternative vehicles. These estimated growth rates will not affect the relative merits of various alternative vehicles, but they will determine *when* we will meet various societal goals, such as reducing GHG emissions to 80 % below 1990 levels or cutting oil consumption such that oil from friendly American nations could supply all US oil demands in some future oil crisis.

Total Number of Light-Duty Vehicles (LDVs) on the Road

The number of LDVs (passenger cars and light-duty trucks) OTR is projected to grow by the DOE's Energy Information Administration from approximately 228 million in 2011 to approximately 291 million by 2040 (Fig. C.1). If we assume a linear extrapolation of the AEO-2014 data through the end of the century, then there would be approximately 429 million LDVs OTR by the end of the century as shown in Fig. C.1, while this linear extrapolation may be too optimistic since it implies some combination of very high population growth, more cars per person (which is already approaching one vehicle per adult in the USA) While this growth rate may not materialize, it should not change the relative merits of various alternative vehicles to reduce pollution and fossil fuel consumption.

C.E. (Sandy) Thomas, *Sustainable Transportation Options for the 21st Century and Beyond*, DOI 10.1007/978-3-319-16832-6

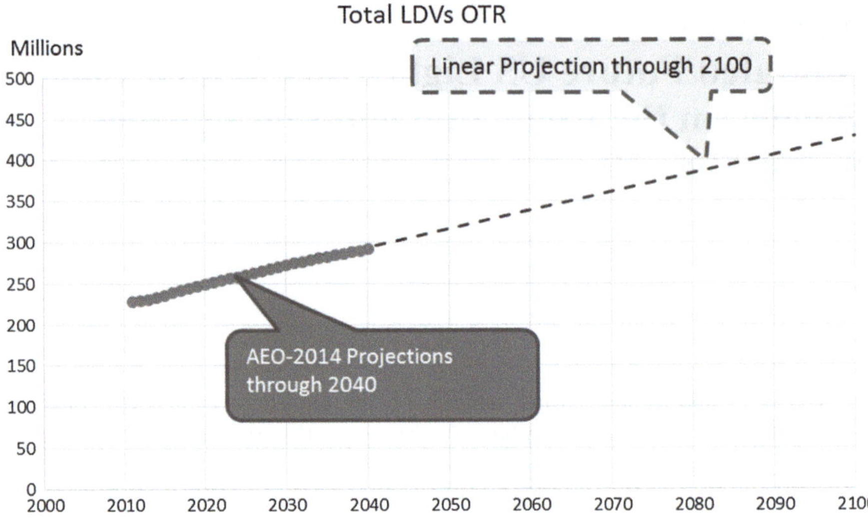

Fig. C.1 Historical and projected number of US LDVs on the road

Annual LDV Travel per Vehicle

To model the GHG emissions and oil consumption over the twenty-first century, we also need to estimate the miles traveled by each vehicle. Figure C.2 shows the historical growth in VMT, the linear extrapolation of those historical data, and two future projections: the EIA's projection from 2011 to 2040, and the linear extrapolation of those AEO-2014 data through the end of the century, which we use

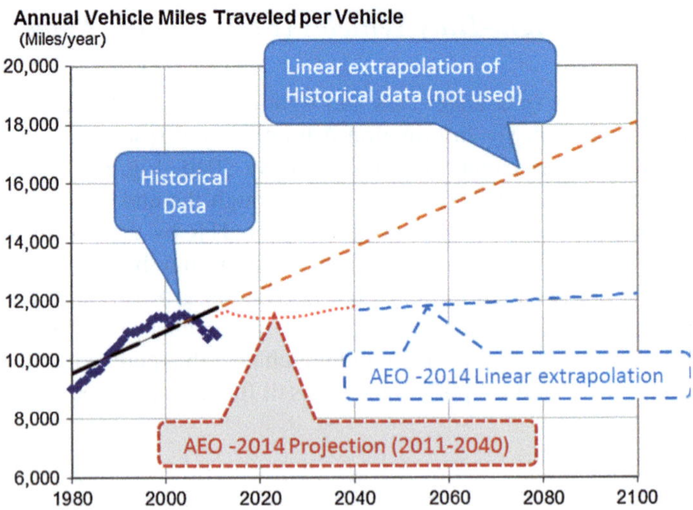

Fig. C.2 Historical and projected US vehicle miles traveled per vehicle

in the model to estimate GHG emissions and oil consumption [C.1]. Thus, this model projects just over 12,000 miles per year per vehicle by 2100, far less than the 18,000 miles per vehicle predicted by a linear extrapolation of historical data. Thus, we assume that the rapid rise in VMT per vehicle in the twentieth century cannot be sustained in the twenty-first century.

Alternative Vehicle Sales Fractions

For each alternative vehicle, we estimate the fraction of new cars sold each year that will be that type of alternative vehicle. All other vehicles sold will either be conventional internal combustion vehicles (ICVs) or another type of AFV that would reduce GHGs or oil consumption.

Gasoline-Powered Hybrid Electric Vehicle (HEV) Sales

For the HEV sales, we already have a multiyear sales history on which to base our projections. As shown in Table C.1, over 3 million HEVs were sold in the USA between 1999 and 2013. More than 70 % of these HEVS were sold by Toyota including 1.54 million of their iconic Prius HEVs[17][C.2].

In this computer model, we start with the historic HEV sales record and project into the future using a logistics function fit to the historic data as shown in Fig. C.4 for the early years and Fig. C.5 over the twenty-first century. The logistics function is given by:

$$F = A/[1 + \exp(G \times (Y - Y_0)] \tag{C.1}$$

where $A = 0.985$, which means that 98.5 % of all LDVs sold will be hybrids by the end of the century; $G = 0.18$, a curve shape factor; and $Y_o = 2032$, the year when half of all LDVs sold are HEVs.

The total number of LDVs sold each year is taken from the AEO-2014 projections as shown in Fig. C.3. We use the Department of Energy's AEO-2014 projections through 2040 [C.1] and then reduce the linear extrapolation of the AEO-2014 data by 85 % between 2040 and 2100 on the assumption that the number of LDVs sold will not maintain the AEO projection through 2100. By 2100, the model therefore projects annual sales of 18.6 million LDVs rather than the 21.9 million annual sales implied by a linear extrapolation of the AEO-2014 data.

Equation C.1 yields the estimated number of HEVs sold each year as shown in Fig. C.4 in the short term and in Fig. C.5 through the twenty-first century. By 2100 in this model, 98.5 % of all LDVs sold are HEVs.

[17]Although the Prius is the most famous HEV, it was not the first HEV sold in the USA. Honda sold 17 of their small insight HEVs in 1999 in the USA before Toyota sold their first 5562 Prius HEVs in 2000.

Table C.1 Number of hybrids sold in the USA between 1999 and 2013

Toyota/Lexus	2,185,696	70.79 %
Honda/Acura	361,559	11.71 %
Ford/Mercury/LIncoln	292,171	9.46 %
Chevy/GMC/Buick/Cadilac	92,039	2.98 %
Hyundai	62,188	2.01 %
Nissan/ Infiniti	39,474	1.28 %
Kia	24,003	0.78 %
Saturn	11,667	0.38 %
VW	6575	0.21 %
Porsche	4272	0.14 %
BMW	3185	0.10 %
Mercedes-Benz	2236	0.07 %
Mazda	1144	0.04 %
Audi	1124	0.04 %
Dodge/Chrysler	88	0.00 %
Totals	3,087,421	100 %

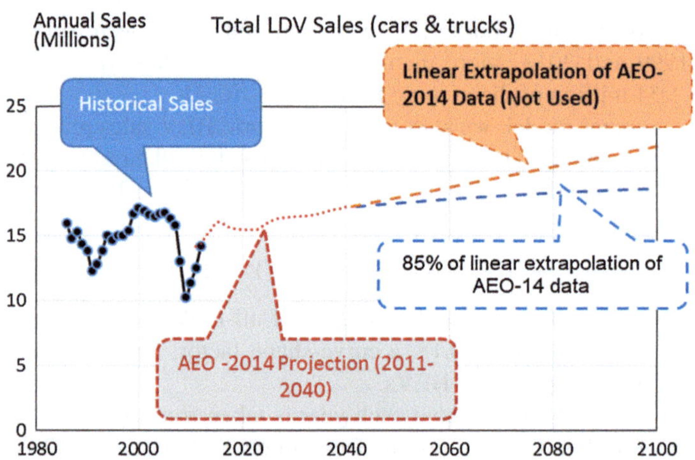

Fig. C.3 Total number of light-duty vehicles (LDVs) (passenger cars and light-duty trucks) in the USA over the twenty-first century

Gasoline-Powered Plug-in Hybrid Electric Vehicle (PHEV) Sales Projections

We have fewer data on PHEV sales, and since sales of the first commercial PHEV, the Chevy Volt only began in 2010. Through the 3rd quarter of 2014, approximately 42,500 PHEVs have been sold in the USA [C.3].[18] Assuming that 4th

[18]Caution is advised with this online "scorecard," since they list non-plug-in pure battery EVs on the same chart as PHEVs.

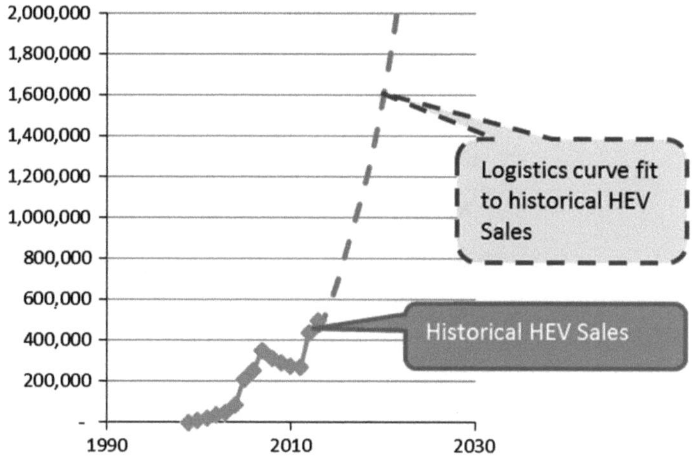

Fig. C.4 Estimated annual US HEV sales in the near term

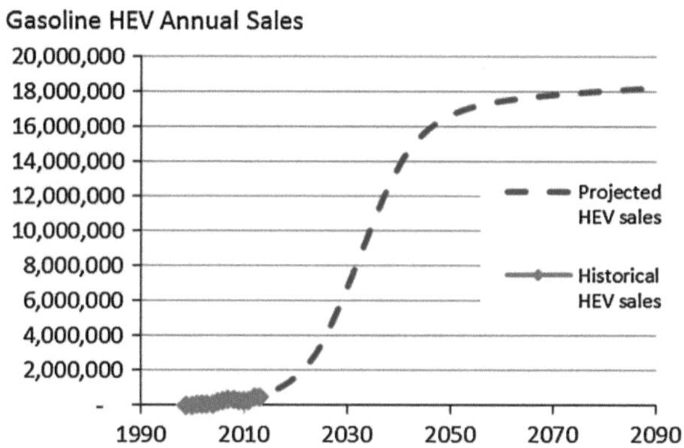

Fig. C.5 Projected annual US HEV sales over the twenty-first century

quarter 2014 PHEV sales equal 3rd quarter sales, then total 2014 PHEV sales would grow to 56,770 vehicles. These historical sales of PHEVs in the USA along with the logistics model curve fit in the near term are shown in Fig. C.6. The logistics curve fit data for the PHEV are:

$$A = 0.985$$
$$G = 0.173, \text{ and}$$
$$Y_o = 2045.$$

The long-term PHEV sales are shown in Fig. C.7 over the twenty-first century.

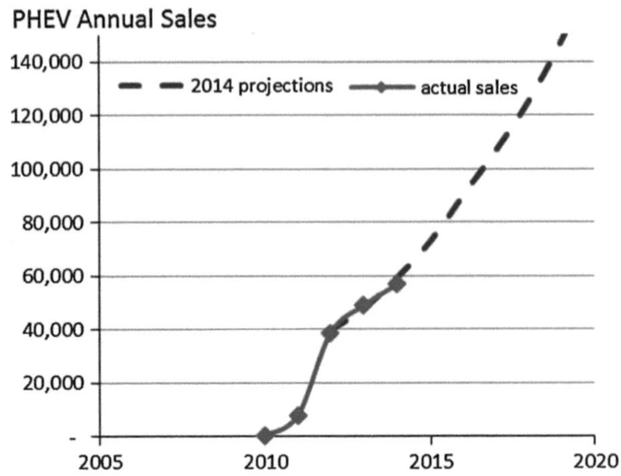

Fig. C.6 Projected near-term PHEV US sales

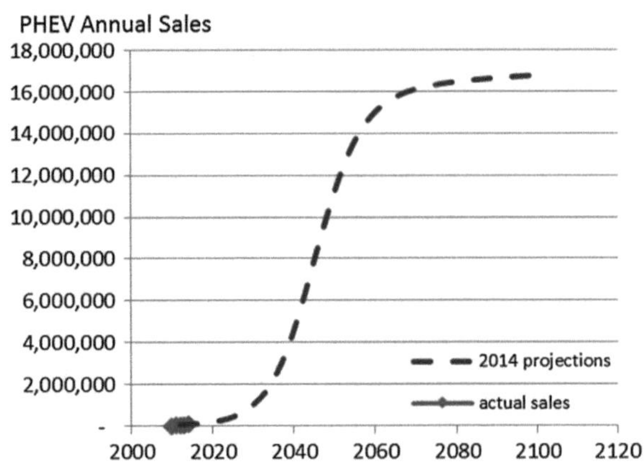

Fig. C.7 Projected US PHEV sales over the twenty-first century

Ethanol-Powered PHEV Sales Projections

No commercial OEM is currently selling ethanol-powered PHEVs, but the transition to such an AFV could be swift since a) the OEMs are producing many "flex-fuel" conventional vehicles that can run on gasoline or mixtures of gasoline and ethanol and b) there are thousands of fuel stations across the USA that sell ethanol/gasoline blends. According to the DOE's Alternative Fuels Data Center, there are 2389 fuel stations in the USA that currently supply E-85, a mixture of 85 % ethanol and 15 % gasoline [C.4].

One major objection to using ethanol is the "food versus fuel" dilemma. Almost all ethanol in the USA is made today by fermenting corn grain. While the ethanol production plants use field corn, and not sweet corn normally consumed by humans, the field corn is fed to cattle and other animals that supply meat for human consumption. In the first decade of the twenty-first century, it was widely expected that ethanol would soon be produced from lignocellulosic materials such as the non-food stalks and roots of corn plants (called "stover") or switchgrass, the native prairie grass that is sometimes used in the farm belt to control runoff of agricultural chemicals.

Lignocellulosic ethanol not only alleviates the "food versus fuel" conundrum, but also substantially reduces greenhouse gas (GHG) emissions compared to ethanol made from corn. As shown in Fig. C.8, ethanol made from corn only reduces GHGs by 17–23 % depending on the corn ethanol plant type, while lignocellulosic ethanol reduces GHGs by 64 % [C.5].

Cellulosic ethanol would also significantly reduce the consumption of fossil fuels for transportation as shown in Fig. C.9 [C.5]. Thus, a gallon of

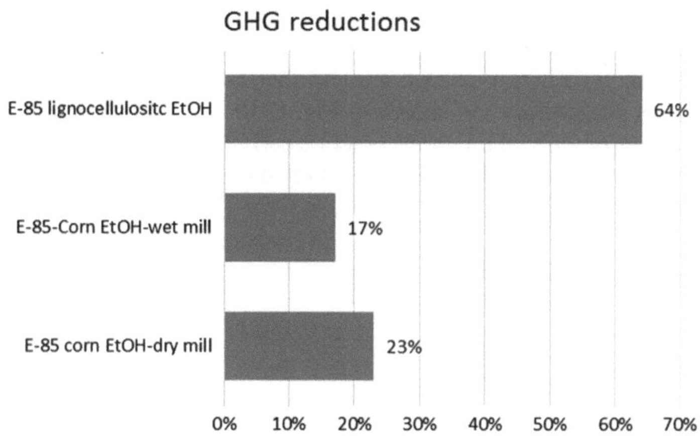

Fig. C.8 Percentage decrease in GHG emissions for E-85 fuel (85 % ethanol and 15 % gasoline)

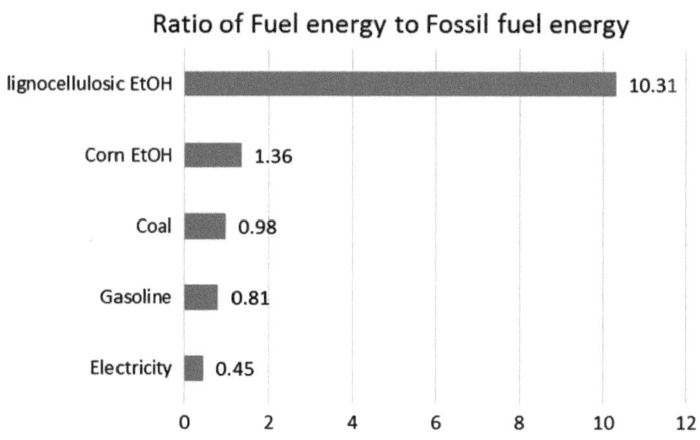

Fig. C.9 Ratio of transportation fuel energy to fossil energy required to make that fuel (large is good!)

lignocellulosic ethanol that contains 76,350 Btu's of energy would require 10.31 times less fossil fuel energy or only 7405 Btu's of fossil fuel energy input, while corn ethanol would require 56,140 Btu's of input fossil fuel energy. Electricity is the worst fuel in this regard, since on the average, electricity contains only 45 % of the fossil energy used to make the electricity from fossil fuels. And gasoline contains only 81 % of the energy in the crude oil used to make the gasoline. Hence, cellulosic ethanol requires much less fossil fuel than any other transportation fuel.

Despite these significant advantages of cellulosic ethanol, the commercial future of this biomass fuel is uncertain as of 2015. Several cellulosic ethanol plants were built in the 2008–2012 time period, but they have not yet proven to be economic. The National Renewable Energy Laboratory [C.6] estimates that the cost of cellulosic ethanol would be approximately $2.05/gallon; since ethanol has 32.8 % lower energy content than gasoline, this would be equivalent to gasoline at approximately $3.11/gallon. For comparison, ethanol made from corn is estimated at a cost between $0.60 and $1.67/gallon.

Nonetheless, for this model, we assume that cellulosic ethanol does become economic over the century. As shown in Fig. C.10, we assume that all ethanol is made from corn initially, the current practice early in the century, shifting to corn stover and to switchgrass, until 96 % of all ethanol is made from these two cellulosic sources by 2100.

The number of PHEVs fueled by E-85 instead of gasoline is governed by another logistics function with the parameters

$$A = 0.95;$$
$$G = 0.26;$$
$$\text{and } Y_0 = 2065$$

as shown in Fig. C.11. These are the fraction of PHEVs fueled by E-85, which leads to the first E-85 PHEVs OTR in 2018, e.g., the product of Fig. C.7, the number of PHEV sales times Fig. C.11 the fraction of PHEVs fueled by E-85.

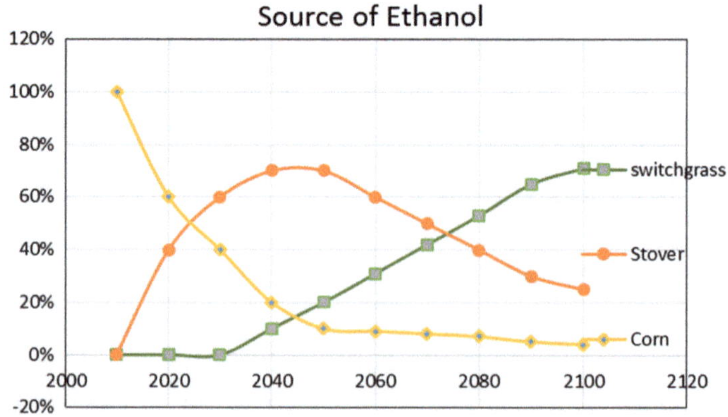

Fig. C.10 Source of fuel ethanol production over the twenty-first century

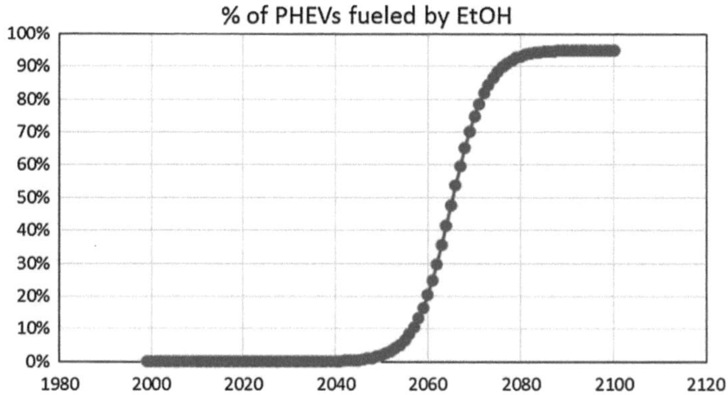

Fig. C.11 Fraction of PHEVs fueled by E-85 (as compared to the total PHEVs from Fig. C.7)

Hydrogen-Powered Fuel Cell Electric Vehicle (FCEV) Sales Projections

Several major automobile companies have vowed to begin commercial production of FCEVs beginning in 2014 or 2015. Several hundred prototype FCEVs were produced for early testing programs, primarily in California, and the California Air Resources Board (CARB) is planning on providing the necessary hydrogen infrastructure so that the auto companies can begin introducing FCEVs as part of the CARB zero-emission vehicle (ZEV) program. The alternative vehicles necessary to meet the CARB goal of cutting GHGs to 80 % below 1990 levels in California by 2050 are shown in Fig. C.12; they are projecting that 87 % of all California cars will have to be electric vehicles (BEVs or FCEVs) by 2050 to meet their GHG reduction targets and that approximately 70 % of all LDVs *OTR* will have to be hydrogen-powered FCEVs as shown in Fig. C.12 taken from a 2012 CARB staff report. [C.7]

The US Department of Energy conducted an in-depth on-road evaluation or "learning demonstration" of 183 FCEVs from four automobile companies[19] and recorded the FCEV performance data for over 500,000 FCEV trips totaling 3.59 million miles, including 35,748 individual hydrogen fueling events without any major safety incident [C.8].

For this model, we used the actual FCEVs used in the learning demonstration program, along with the FCEVs that are projected for the ZEV program in California as summarized in Fig. C.13 for the near term. For the longer term, we used another logistics curve with these parameters

$$A = 0.99;$$
$$G = 0.28$$
$$Y_o = 2038$$

as shown in Fig. C.13 for the near term and in Fig. C.14 for the entire century.

[19]Daimler, Ford, GM, and Hyundai-Kai.

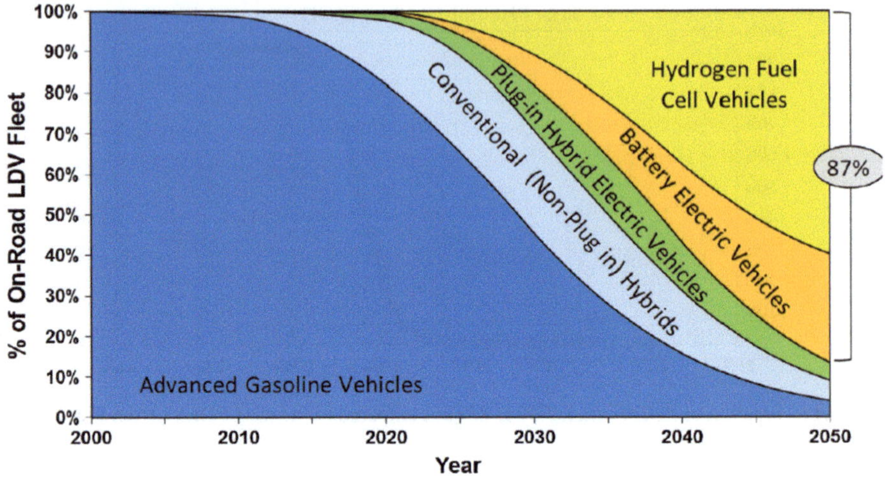

Fig. C.12 Estimated alternative vehicles on the road required to meet the California goal of cutting GHGs to 80 % below 1990 levels in California by 2050

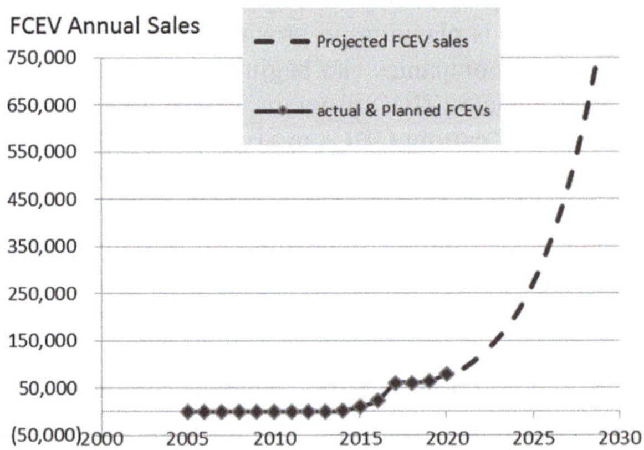

Fig. C.13 Actual, planned, and projected FCEV sales through 2030

Battery-Powered Electric Vehicles (BEVs) Sales Projections

The sales of BEVs will be limited by several factors: short range between battery charges, long charging times, heavy vehicles even to meet minimal ranges, and the lack of a place to recharge batteries for many drivers. The range limitations of BEVs caused by mass compounding are discussed in detail in Appendix A, and the impacts of both range limitations and limited off-street parking are summarized

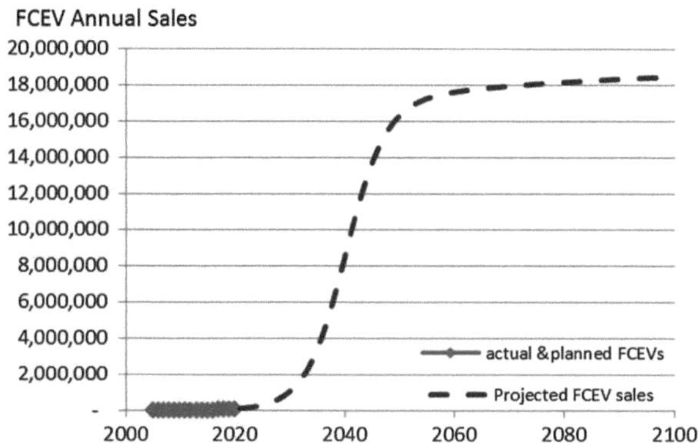

Fig. C.14 Projected FCEV sales over the twenty-first century

Table C.2 Historical US BEV annual sales

BEV	2010	2011	2012	2013	2014	Totals
Nissan Leaf	19	9674	9819	23,094	30,200	72,806
Ford Focus Electric			685	1738	1964	4387
Mitsubishi i-MiEV		80	588	1029	196	1893
Rav-4 EV			192	1096	1184	2472
Chevy spark EV				539	1145	1316
Honda Fit			93	569	407	1069
Tesla Model S (US)			2650	17,650	17,300	37,600
Cadillac ELR				6	1310	1316
Porsche Panameria S-E				86	879	965
Fiat 500e				2310	5132	7442
Smart ED				923	2594	3517
Mercedes B-Class ED					774	774
BMW i8					555	555
Total BEVs	19	9754	14,027	49,040	69,732	142,572

in Appendix B. Carnegie Mellon University estimates that only 56 % of drivers have access to garages or off-street parking where BEVs could be charged [C.9]; many drivers live in large metropolitan areas where they have to search for a parking place every night and therefore do not have access to a fixed place to routinely charge their car batteries.

The historical BEV sales in the USA are summarized in Table C.2,[20] starting with 19 Nissan Leaf BEVs sold in 2010, accelerating to 57,955 BEVs estimated for 2014.[21]

[20]Data were only available through the end of the third quarter of 2014; we have assumed here that fourth quarter 2014 BEV sales are equal to third quarter BEV sales for all BEV models.

[21]The Tesla BEV sales are approximations, since Tesla is the only automaker that does not report annual sales for the USA. These Tesla numbers may also include some BEV sold overseas.

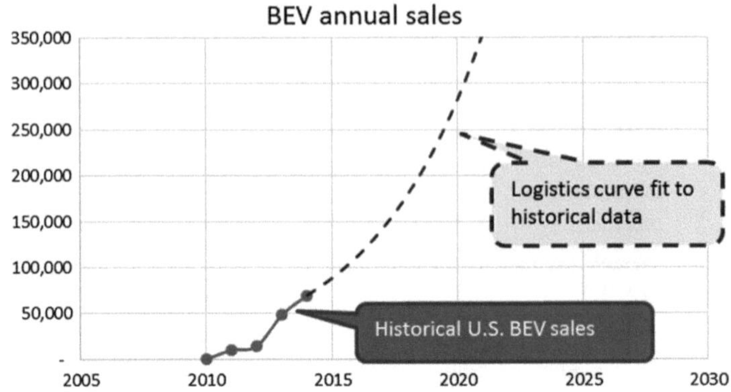

Fig. C.15 Near-term US BEV annual sales logistics projection matched to historical US BEV annual sales

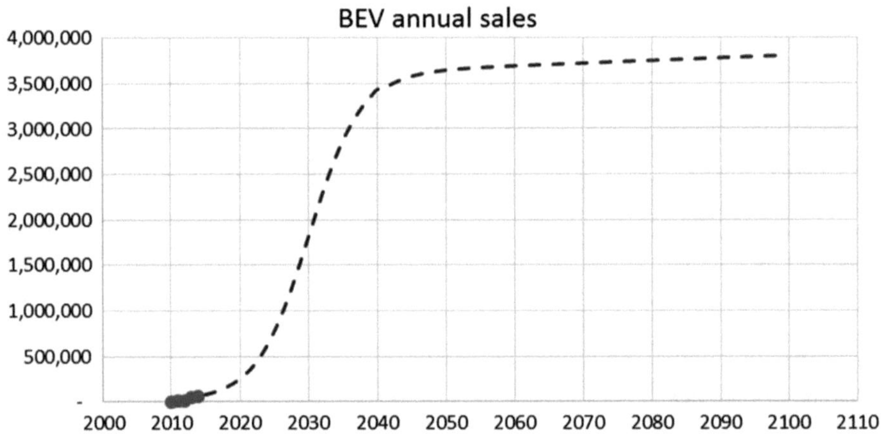

Fig. C.16 Long-term U S BEV sales history assuming a 31.1 % maximum market penetration and a logistics curve fit to historical sales data

For the base case of 30 % BEV maximum market penetration, the historical and projected BEV US sales are shown in Fig. C.15 for the near term and Fig. C.16 for the twenty-first century. The logistics function parameters for this curve fit are:

$$A = 0.311$$
$$G = 0.245$$
$$\text{and } Y_o = 2030$$

References

[C.1] The U.S. Department of Energy, Energy Information Administration (EIA), Annual Energy Outlook for 2014 (AEO-2014), available at: http://www.eia.gov/oiaf/aeo/tablebrowser/

[C.2] "Hybrid electric vehicles in the United States," available at: http://www.ask.com/wiki/Hybrid_electric_vehicles_in_the_United_States?o=2801&qsrc=999&ad=doubleDown&an=apn&ap=ask.com#Historical_sales_1999-2013

[C.3] "Inside EVs: Monthly Plug-in sales Scorecard," available at: http://insideevs.com/monthly-plug-in-sales-scorecard/

[C.4] The U.S. Department of Energy's Alternative Fuels Data Center, run by the National Renewable Energy Laboratory, available at: http://www.afdc.energy.gov/fuels/ethanol_locations.html

[C.5] M. Wang, "Updated Energy and Greenhouse Gas Emissions results of fuel ethanol," Presented to the 15th International Symposium on alcohol fuels, San Diego, California, September 2005.

[C.6] A Dutta et al., "Process Design and Economics for Conversion of Lignocellulosic Biomass to Ethanol," National Renewable Energy Laboratory report NREL/TP-5200-51400. Mau 2011.

[C.7] California Air Resources Board Staff report: "an initial statement of reasons: Advanced Clean Cars; 2012 Proposed Amendments to the California Zero Emission Vehicle program regulations," released December 7, 2011. Available at: http://www.arb.ca.gov/regact/2012/zev2012/zevisor.pdf

[C.8] K. Wipke et al., "National Fuel Cell Electric Vehicle Learning Demonstration Final Report," National Renewable Energy Laboratory Report NREL/TP-5600-54860, July 2012, available at: http://www.nrel.gov/hydrogen/pdfs/54860.pdf

[C.9] E.J. Traut et al., "U.S. residential charging potential for electric vehicles," Carnegie Mellon University, *Transportation Research Part D* 25 (2013) 139-145. Available at: http://www.cmu.edu/me/ddl/publications/2013-TRD-Traut-etal-Residential-EV-Charging.pdf

Appendix D
Marginal Grid Mixes

Introduction

When calculating the greenhouse gas (GHG) emissions from electrical power plants used to recharge batteries on BEVs or plug-in hybrid electric vehicles (PHEVs), some analysts use the *average* grid mix for the local utility. But this often underestimates the actual GHGs generated when batteries are charged. The correct metric is the *marginal* grid mix which takes into account the generator that is "on the margin" ready to be turned on when a new load such as a BEV or PHEV is added to the grid load.

Marginal Grid Mix Illustration

Consider a hypothetical utility that has only two generators: a coal power plant and a hydroelectric power plant, each serving half the peak load. Assume further that during the evening hours, the utility load is half the total capacity of these two generators. To minimize their costs, this utility would turn on the hydroelectric plant first, since it has the lowest operating cost. In the evening, with a 50 % load, the hydroelectric plant would supply all power.

If a new load (such as charging the batteries on BEVs and PHEVs) is added, then the utility would have to turn on the coal plant to supply this extra load.[22] The coal plant would be on the margin.

The average emissions from these two power plants are approximately 500 g of CO_2 per kWh generated, the average of 1000 g/kWh for the coal plant and zero for the hydroelectric plant.

[22]In reality, a coal plant cannot be turned on and off quickly, since it takes many hours to heat up a coal plant. So in this example, the coal plant would have to on all the time in a "hot standby" mode, ready to ramp up whenever the load required more power than the hydroelectric plant could deliver.

C.E. (Sandy) Thomas, *Sustainable Transportation Options for the 21st Century and Beyond*, DOI 10.1007/978-3-319-16832-6

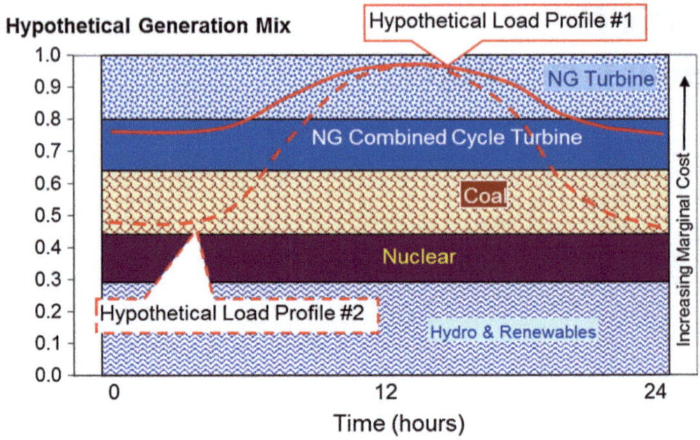

Fig. D.1 Utility generator mix arranged with highest operating cost generators on top, and lowest operating cost generators on the bottom along with two hypothetical load profiles

An analyst using average emissions would therefore assign a GHG value of 500 g/kWh to a BEV or PHEV added to this grid, when in fact adding the BEV or PHEV would generate 1,000 g/kWh, the GHG emissions from the coal plant on the margin. In this hypothetical case, the average grid mix result would be *twice* the correct result from the marginal grid mix.

As stated in the Greenhouse gas protocol [D.1]:

> An average emission rate is easy to calculate, but it provides only a rough approximation of marginal displaced emissions. A simple average emission rate may be necessary in situations where data are not available to perform a marginal rate methods. Because calculating a simple average is *significantly less precise* than other methods, however, it should only be used where other methods are not practicable.

Of course actual utility grid mixes are much more complicated than the hypothetical two-plant utility described above. Figure D.1 shows a more realistic mix of generators for a utility that represents the average generator sources of all US utilities. The generators are arranged in increasing operating cost from bottom to top. Thus, the generators with the lowest operating costs[23] (Hydro, renewables, and nuclear) are shown in the bottom of the chart. Having low operating costs, these are the generators that are generally turned on first and run as baseload. Above these generators come generators with higher operating costs, such as coal plants, then natural gas combined cycle (NG CC) plants,[24] and finally the most expensive to operate natural gas (single cycle) combustion turbines (NG CT). Also shown in Fig. D.1 are two hypothetical utility load profiles. The intersection of these load

[23]Note that this metric ignores the capital cost of these generators. Thus nuclear plants are very expensive to build, but have low operating costs once they are constructed.

[24]Combined cycle plants have higher efficiency than single-cycle turbines or combustion turbines, so the combined cycle plants use less natural gas and have lower operating costs.

profiles with the stacked utility generator profiles defines the likely generators that will supply those loads—the marginal grid mixes. For load profile #1, all the electricity would be generated from natural gas plants: combined cycle plants around noon, and lower efficiency combustion turbines in the evening and late morning and afternoon hours.

Load profile #2 dips into the coal plant regime from early afternoon, through the night, until mid-morning, so the coal plants would supply a large fraction of the electricity for load profile #2, so GHGs would be much higher with load #2 compared to load #1. For BEVs and PHEVs charging at night, load profile #2 will generate more GHGs than load profile[25] #1.

Actual US Marginal Grid Mixes

As stated in the GHG protocol above, calculating the marginal grid mix is much more complex than calculating the average grid mix, since calculating the marginal grid mix requires an hour-by-hour load profile for each utility system to determine which generators are on the margin, information that is not readily available. Fortunately, the scientists and engineers at the Oak Ridge National Laboratory have made such complex calculations for each of the major US electricity regions (See Fig. D.2 for map of the 2007 EIA regions) using data from the DOE's 2007 Annual Energy Outlook [D.2]; we use their marginal grid mix data (top half of Table D.1) for 2020 in this model; we entered these utility marginal grid mix data into the GREET model to arrive at GHG emissions and petroleum consumption for both average and marginal grid mixes. Note that nuclear power (with zero GHG emissions) makes up more than 18 % of the average LDV-weighted grid mixes,[26] but the electricity from these nuclear plants is never on the margin in any region of the country and new loads such as BEVs would never lead to an increase in nuclear power production; therefore, any analysis that includes nuclear power in the calculation of GHG emissions from electric vehicles is flawed.

The differences between marginal and average grid mixes show up predominantly with BEVs and to a lesser extent with PHEVs. Table D.2 shows the results of BEV GHG emissions for the 14 grid regions mentioned above. In 8 of the 14 regions, GHGs for BEVs increase using the marginal grid mix compared to the average grid mix, and GHGs are lower for marginal average grid mixes in the other 6 regions.

[25]Note that the zero GHG generators (hydro, renewables, and nuclear) are never on the margin in this simplified example and are almost always run as baseload as the lowest operating cost generators; so a high percentage of zero-GHG generators at a given utility do not necessarily reduce the actual GHG emissions from these utilities when new loads such as BEVs or PHEVs are added.

[26]Nuclear power contributed more than 20 % of all US power in this time period, but the weighting by LDV in each region results in the 18.3 % estimate, meaning that those regions with high concentrations of registered LDVs tend to have lower concentration of nuclear power plants.

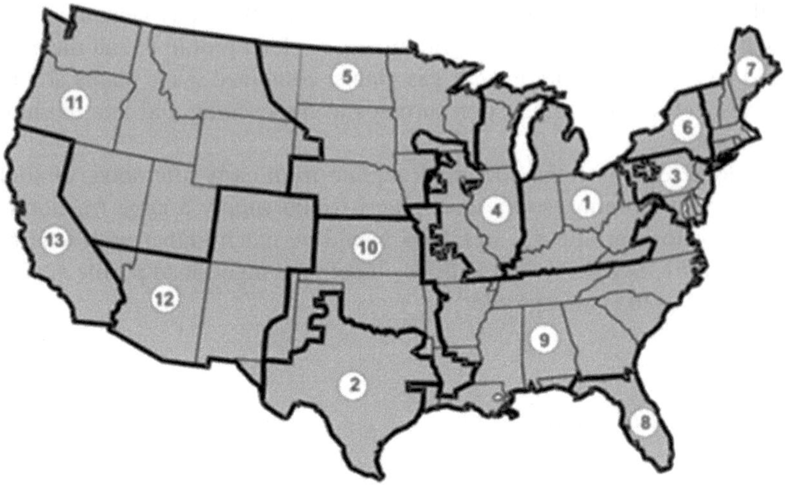

Figure ES-1. Regions as specified by EIA for *Auuual Euergy Outlook 2007*:

1. ECAR—East Central Area Reliability Coordination Agreement
2. ERCOT — Electric Reliability Council of Texas
3. MAAC — Mid-Atlantic Area Council
4. MAIN — Mid-America Interconnected Network
5. MAPP— Mid-Continent Area Power Pool
6. NPCC -NY — Northeast Power Coordinating Council / NY
7. NPCC -NE — Northeast Power Coordinating Council / New England
8. FRCC — Florida Reliability Coordinating Council;
9. SERC — Southeastern Electric Reliability Council
10. SPP — Southwest Power Pool
11. WECC-NW — Western Electricity Coordinating Council / Northwest Power Pool Area
12. WECC-RMP/ANM — Western Electricity Coordinating Council / Rocky Mountain Power Area and Arizona-New Mexico-Southern Nevada Power Area
13. WECC-CA — Western Electricity Coordinating Council / California

Fig. D.2 Electricity regions used in the EIA's 2007 Annual Energy Outlook

The last row in Table D.2 also shows the percentage of LEVs in each region.

The results of calculating the GHG emissions for all alternative fueled vehicles are shown in Table D.3. The results are obviously identical for AFVs that do not use any grid electricity. The LDV-weighted GHG estimate from the marginal grid data for BEVs is 9.86 % greater than the GHGs calculated using the average grid mixes in the 14 regions. The GHGs for PHEVs using the marginal grid mixes are from 3.6 to 5.3 % greater than the GHGs calculated using the LDV-weighted average grid mix.

Table D.1 Marginal and average electrical grid mixes from the Oak Ridge National Laboratory (first 13 regions) (Region 14 (Alaska and Hawaii) modeled manually by author)

Marginal grid mixes							
	East Central Area Reliability Coordination Agreement—01 (%)	Electric Reliability Council of Texas—02 (%)	Mid-Atlantic Area Council—03 (%)	Mid-America Area Interconnected Network—04 (%)	Mid-Continent Area Power Pool—05	Northeast Power Coordinating Council/New York—06	Northeast Power Coordinating Council/New England—07
Coal	55.9	0.0	25.5	62.0	22.7	4.0	14.8
Oil	0.3	0.0	11.0	0.0	2.1	13.6	20.7
Gas ST	0.3	24.5	0.2	0.5	0.2	12.8	0.8
Gas CC	20.0	52.5	57.4	9.2	30.1	52.5	49.4
Gas CT	19.9	22.4	8.1	28.2	36.8	16.9	5.4
Total natural gas	42.9	99.4	65.8	37.9	67.1	82.3	55.6
Renew	0.8	0.0	0.0	0.0	0.0	0.0	0.0
Unserved	0.0	0.1	0.1	0.0	0.0	0.0	0.0
Nuclear	0.0	0.0	0.0	0.0	0.0	0.0	0.0
LDV % per region	13.7	5.9	9.1	8.8	3.6	5.4	5.2

(continued)

Table D.1 (continued)

	Florida Reliability Coordinating Council—08	Southeastern Electric Reliability Council—09	Southwest Power Pool—10	WECC/ Northwest Power Pool Area—11	WECC/Rocky Mountain, Arizona, New Mexico, S. Nevada Power Area—12	WECC/ California—13	Alaska and Hawaii—14	LDV-wgt'd Marginal Grid Mix for 2020
Coal	0.3	33.3	5.1	0.0	7.2	0.0	4.2	23.4
Oil	16.3	0.1	2.3	0.0	0.0	0.1	70.1	4.4
Gas ST	11.0	0.3	13.3	15.9	1.9	21.7	6.1	7.6
Gas CC	51.1	61.9	75.1	73.4	88.2	60.5	14.2	48.9
Gas CT	18.6	2.9	4.1	10.6	2.6	16.9	6.1	14.4
Total natural gas	80.7	65.1	92.5	100.0	92.6	99.2	26.4	71.0
Renew	0.6	1.5	0.0	0.0	0.2	0.5	0.0	0.5
Unserved	0.6	0.0	0.0	0.0	0.0	0.2	0.0	0.1
Nuclear	0.8	0.0	0.0	0.0	0.0	0.0	0.0	0.0
LDV % per region	5.5	17.1	2.9	5.1	3.0	13.4	0.6	

(continued)

Table D.1 (continued)

Average grid mixes	East Central Area Reliability Coordination Agreement—01 (%)	Electric Reliability Council of Texas—02 (%)	Mid-Atlantic Area Council—03 (%)	Mid-America Area Interconnected Network—04 (%)	Mid-Continent Area Power Pool—05	Northeast Power Coordinating Council/New York—06	Northeast Power Coordinating Council/New England—07
Coal	78.3	32.1	49.3	53.1	73.3	13.1	11.5
Oil	0.0	0.0	1.0	0.0	0.1	1.1	2.1
Gas ST	0.2	4.2	0.0	0.0	0.0	1.0	0.0
Gas CC	10.5	45.5	14.2	12.3	4.0	39.7	55.3
Gas CT	1.1	4.7	0.4	1.9	2.6	1.0	0.4
Total natural gas	11.8	54.4	14.7	14.2	6.6	41.7	55.8
Renew	1.1	0.4	3.4	1.5	5.9	17.5	10.9
Unserved	0.0	0.0	0.0	0.0	0.0	0.0	0.0
Nuclear	8.8	11.0	31.3	11.4	11.4	26.1	19.7

(continued)

Table D.1 (continued)

	Florida Reliability Coordinating Council—08	Southeastern Electric Reliability Council—09	Southwest Power Pool—10	WECC/ Northwest Power Pool Area—11	WECC/Rocky Mountain, Arizona, New Mexico, S. Nevada Power Area—12	WECC/ California—13	Alaska and Hawaii—14	LDV-wgt'd Marginal Grid Mix for 2020
Coal	34.7	51.9	67.8	35.0	66.3	30.1	12.2	46.6
Oil	3.2	0.3	0.3	0.0	0.0	0.0	56.0	0.9
Gas ST	3.5	2.9	2.0	0.3	0.7	5.0	6.4	1.8
Gas CC	47.0	8.4	20.9	16.8	17.3	27.3	7.9	21.6
Gas CT	4.4	0.5	0.7	0.6	0.5	4.4	3.6	1.8
Total natural gas	54.9	11.9	23.7	17.8	18.5	36.7	17.9	25.2
Renew	2.3	3.9	2.4	41.9	4.6	18.4	13.9	8.1
Unserved	0.0	0.0	0.0	0.0	0.0	0.0	0.0	0.0
Nuclear	4.7	32.0	3.9	2.9	10.1	12.0	0.0	18.2

Table D.2 GHG emissions in grams/mile for BEVs using average and marginal grid mixes, % change of marginal compared to average grid mix calculations, and % of LDVs in each of the 14 US geographic regions

Region	1	2	3	4	5	6	7	8	9	10	11	12	13	14	Wgt'd Ave.
BEV Marginal GHGs	407	291	337	434	327	313	307	304	312	266	260	245	277	443	326.8
BEV Average GHGs	415	292	284	299	382	168	192	320	292	392	217	371	246	392	297.5
delta	−1.9	−0.6	18.6	45.2	−14.2	86.3	59.6	−4.8	7.0	−32.1	20.1	−33.8	12.8	12.9	14.5
of LDVs	13.7	5.9	9.1	8.8	3.6	5.5	5.3	5.6	17.2	3.0	5.2	3.1	13.5	0.6	

Table D.3 LDV-weighted GHG emissions comparing average with marginal grid mixes and % increase of marginal GHGs relative to average GHGs

	GHG emissions (g/mile)		
	Marginal	Average	Marg/Ave (%)
Gasoline ICV	549.8	549.8	0.00
E-85 ICV	358.8	358.8	0.00
Gasoline HEV	339.0	339.0	0.00
NGV	488.1	488.1	0.00
NG HEV	309.8	309.8	0.00
Diesel ICV	474.4	474.4	0.00
Diesel HEV	307.9	307.9	0.00
E-85 HEV	309.8	309.8	0.00
BEV	326.6	297.3	9.86
FCEV	256.2	256.2	0.00
NG PHEV-40	300.1	297.3	0.93
FC PHEV-40	290.6	275.8	5.34
EtOH PHEV-40	361.9	349.4	3.57
Gasoline PHEV-40	285.9	273.5	4.56

References

[D.1] Breoekhoff, D. "The greenhouse gas protocol: guidelines for quantifying GHG reductions from grid-connected electricity projects," World Resources Council and the World Business Council for sustainable Development. Available at: http://pdf.wri.org/GHGProtocol-Electricity.pdf August 2008.

[D.2] Hadley S, Tsvetkova A. "Potential impact of PHEVS on regional power generation," Report # ORCL/Tm-2007/150. Available at: http://www.ornl.gov/info/ornlreview/v41_1_08/regional_phev_analysis.pdf, January 2008.

Appendix E
Capital Cost of Equipment to Produce Hydrogen, Heat, and Electricity at Wastewater Treatment Plants

Introduction

One excellent source of hydrogen is from waste processed at municipal wastewater treatment plants (WWTPs). Several of these treatment plants are found in every major metropolitan area, and some are already equipped with anaerobic digesters (AD) that use microorganisms to break down the solid waste in the absence of oxygen to produce methane (CH_4), the main ingredient in natural gas, and a superb source of hydrogen.

The key components of a wastewater treatment plant hydrogen tri-generation[27] system are illustrated in Fig. E.1 (heat exchangers not shown that would be used to capture the waste heat from the SMR and fuel cell systems to supply heat to the digester tanks).

Wastewater from the sewage system is concentrated in settling tanks, and the remaining sludge is placed into a set of anaerobic digester tanks. Microbes then convert part of the sewage waste into digester gas (primarily methane and carbon dioxide) in the absence of oxygen. A gas cleanup system then removes most of the carbon dioxide, hydrogen sulfide, and siloxides, leaving a gas made up primarily of methane. This methane is then fed to a steam methane reformer (SMR) that converts the methane and water to hydrogen, some carbon monoxide (CO), and carbon dioxide (CO_2):

$$3H_2O + 2CH_4 = CO + CO_2 + 7H_2 \tag{E.1}$$

A water gas shift reactor then converts most of the residual CO into more hydrogen and more CO_2:

$$H_2O + CO = H_2 + CO_2 \tag{E.2}$$

For a net reaction of:

$$4H_2O + 2CH_4 = 2CO_2 + 8H_2 \tag{E.3}$$

[27]Tri-generation refers to the three products from this plant: hydrogen, heat, and electricity.

C.E. (Sandy) Thomas, *Sustainable Transportation Options for the 21st Century and Beyond*, DOI 10.1007/978-3-319-16832-6

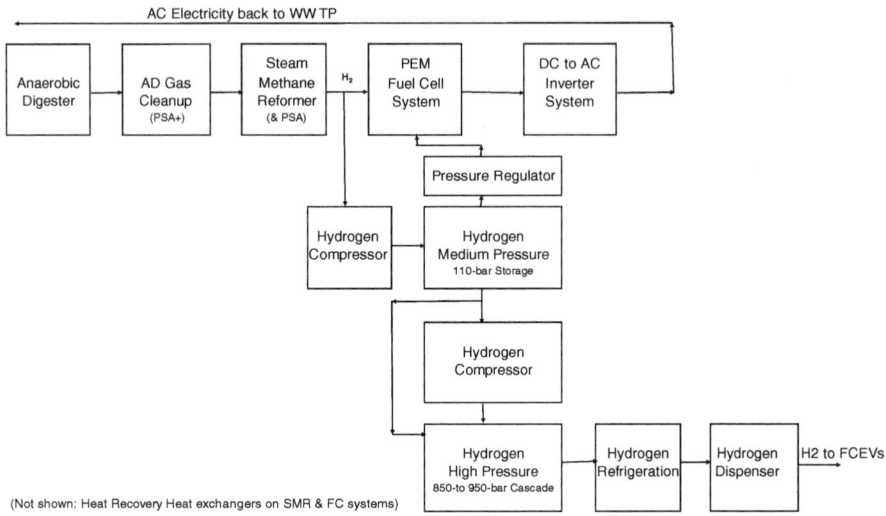

Fig. E.1 Block diagram of a hydrogen tri-generation system at a waste water treatment plant (heat recovery not shown)

So half the hydrogen comes from water and half from methane (8 hydrogen atoms each).

Finally, a pressure swing adsorption (PSA) system produces a very pure stream of hydrogen suitable for the stationary fuel cell system and the fuel cells onboard FCEVs, and the residual CO_2 is released.

This ultra-pure hydrogen is then compressed and stored in a set of storage tanks at a nominal pressure in the range of 110 bar; this medium-pressure stored hydrogen could then be used to fuel the stationary fuel cell during a power outage or during plant downtime for maintenance, providing a very stable supply of electricity for the WWTP.[28]

To fill the hydrogen tanks on the FCEVs, a second compressor[29] raises the pressure of some hydrogen to the range between 850 and 900 bar that is stored in a multitank cascade to fill the 700-bar FCEV car tanks.[30] In the process of filling the high-pressure tanks, the hydrogen is heated, which reduces the amount of hydrogen that can be stored in the car tanks. To assure a full tank of hydrogen, the hydrogen is cooled prior to filling the car tanks.

[28]However, we have not included the cost of this backup electricity system in this analysis.

[29]Alternately, one compressor may have two or more stages, with the first stage compressing the hydrogen from the PSA to 110 bar and the second stage to boost the pressure to 850–900 bar for vehicle fueling.

[30]As indicated in the diagram, some of the 110-bar hydrogen is used initially to fill nearly empty car tanks, and then, higher and higher pressures are added until the car tanks are full at 700-bar pressure.

Table E.1 Available WWTPs in Los Angeles and Orange Counties, and the number of LDVs and population in these two counties, along with the number of FCEVs that could be supported by these 34 WWTPs

	# of WWTPs	Available H2 (tonnes/year)	# of FCEVs supported (M)	Population (millions)	Registered LDVs (M)	LDVs/WWTP
Los Angeles	23	10,000	57.51	9.82	7.2	311,261
Orange County	11	5400	31.06	3.01	2.4	221,727
LA areaTotals	34	15,400	88.57	12.83	9.6	282,294

Available WWTPs in Los Angeles Area

The National Renewable Energy Laboratory has analyzed the number of available WWTPs in the Los Angeles area (LA and Orange Counties).[31] As shown in Table E.1, there are approximately 34 available WWTPs in these two counties that do not already utilize some or all of their biogas to produce either heat or electricity. These 34 plants could produce enough hydrogen to support more than 88,000 FCEVs. There are 9.6 million registered LDVs in these two counties or an average of 282,294 LDVs per WWTP.

One WWTP in Tulare, California, is already producing 1.2 MW of electricity with a molten carbonate fuel cell (MCFC) system supplied by Fuel Cell Energy.[32] The Tulare plant processes 11.7 million gallons of waste per day (MGD), with 4.4 MGD from domestic waste from the city's population of 59,278, and 7.1 MGD from industrial waste. This WWTP requires 2.7 MW of electrical power, so the 1.2 MW of fuel cell power from waste satisfies 45 % of the electrical demand to run the WWTP.

FCEVs Projected for the LA Area

The projected number of FCEVs sold nationwide is derived in Appendix C. The resulting cumulative number of FCEVs OTR per wastewater treatment plant in the Los Angeles area based on this national average FCEV density is shown in Fig. E.2. The number of projected FCEVs is quite small in the early years of market

[31]G. Saur & A. Milbrandt, "Renewable hydrogen potential from biogas in the United States," National Renewable Energy Laboratory Report NRL/TP-5400-60283, July 2014, available at: http://www.nrel.gov/docs/fy14osti/60283.pdf.

[32]Fuel cells 2000 case study, "Fuel cell system turns waste into electricity at the Tulare wastewater treatment plant" available at http://www.fuelcells.org/uploads/TulareCaseStudy.pdf.

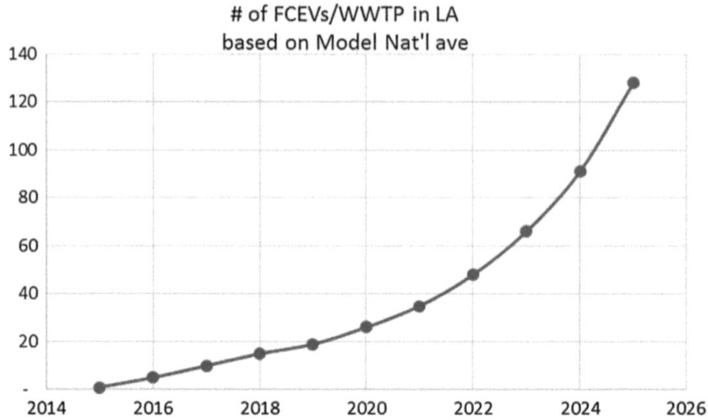

Fig. E.2 Projected number of FCEVs per WWTP in California based on the conservative (low) national average FCEV density in the model

Table E.2 Projected number of FCEVs in the 2017–2020 time period

	California	LA area
2017	6650	2178
2020	18,465	6048

penetration. For example, the model predicts a national average of only 10 FCEVs OTR per WWTP in 2017 and 26 in 2020, based on the number of LDVs in the LA area.

But the expected number of FCEVs in California is much higher as shown in Table E.2 for two LA area counties.[33] The CARB[34] is estimating that there will be 6,650 and 18,465 FCEVs OTR by 2017 and 2020, or 2,178 and 6,048 FCEVs in the LA area based on LDV density in this area.[35]

On the average, the number of FCEVs projected in the LA area per WWTP is 225 times[36] the national average as shown in Fig. E.3, which shows the CARB estimates for 2017 and 2020 superimposed on the data national average data from Fig. E.2 multiplied by 225. So the density of FCEVs relative to LDVs in the LA area will be 225 times greater than the national average during the early years of FCEV deployment, due to the State of California actively promoting hydrogen fueling stations and FCEVs through their ZEV program.

[33]Los Angeles and Orange counties.

[34]"An annual evaluation of fuel cell vehicle deployment and hydrogen fuel station network development," Air Resources Board of the California Environmental Protection Agency, June 2014, available at http://www.arb.ca.gov/msprog/zevprog/ab8/ab8_report_final_june2014.pdf.

[35]The LA area is home to approximately 32.76 % of all California vehicles (9.6 out of 29.3 million vehicles).

[36]From a national average of 10 FCEVs in 2017 to 2178 in 2017 and from 26 to 6048 in 2020.

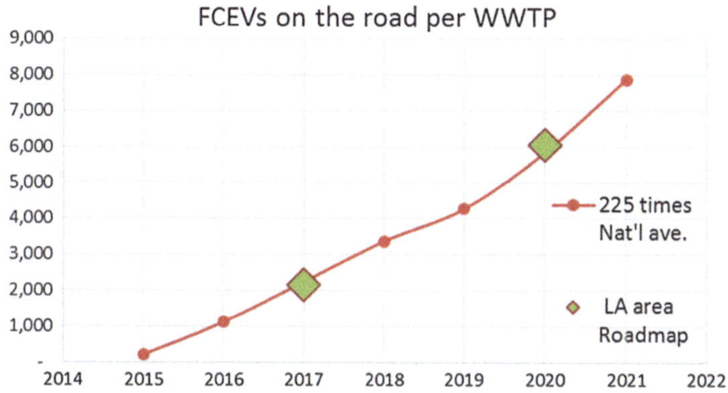

Fig. E.3 Number of FCEVs per WWTP from the CARB projections compared with 225 times the national average projection from Fig. E.2

WWTP Tri-Gen Return on Investment

The internal rate of return (IRR)[37] or return on investment in a tri-generation system will depend on several variables including:

- The capacity of the stationary fuel cell system in terms of electrical output in kW.
- The demand for vehicle hydrogen which will depend on the number of FCEVs in the vicinity of the WWTP.
- The local price of fuels such as natural gas, electricity, and gasoline; the price of natural gas will determine the avoided cost of heating by using the waste heat from the SMR and stationary fuel cell system; the price of electricity will determine the cost savings from generating electricity onsite instead of buying it from the local utility, and the price of gasoline will determine the allowable price of hydrogen for FCEVs.[38]
- The installed cost for all systems.
- The efficiencies of the various systems.

[37]Before taxes and current dollars.

[38]In the model, we assume that the price of hydrogen is set to a value such that the owner of a FCEV pays 10 % *less* per mile (a variable) for vehicle fuel than he or she would pay for gasoline in a conventional vehicle to travel the same distance.

Fuel Cell Sizing

An existing hydrogen Tri-Gen system at the Orange County Sanitation District WWTP in Fountain Valley California produces 250 kW of electrical power,[39] but we have assumed here that a set of automotive or bus style proton exchange membrane (PEM) fuel cell systems produces 1.3 MW (similar to the 1.2 MW fuel cell system at the Tulare WWTP described above) as a base case to maximize the savings by avoiding the costs of electricity consumed on-site. We are postulating the use of stationary PEM fuel cells produced by or for automobile and bus manufacturers as they ramp up production of FCEVs. Specifically, Brian James of Strategic Analysis, Inc. of Arlington, Virginia, recommends using bus PEM fuel cell stacks for this stationary application, since bus fuel cells are typically more durable due to higher catalyst loadings (0.4 mg/of platinum per cm^2 compared to 0.15 mg of Pt/cm^2 for automotive stacks), and bus FC stacks typically operate at lower pressure (1.8 vs. 2.5 atmospheres).[40] The bus PEM Fuel cell system analyzed in detail by James et al. has a power level of 160 kW in two stacks, so 8 or 9 bus FC systems would be used in this stationary 1.3 MW application.

SMR Sizing

The SMR must produce hydrogen for both the stationary fuel cell system and dispensing hydrogen to the FCEVs.

Hydrogen Required for the 1.3 MW Electrical Load

The stationary fuel cell will produce $1300 \times 24 = 31{,}200$ kWh per day. Assuming a FC efficiency of 48 %, then the hydrogen required must have an energy content of:

31,200 kWh/0.48 = 65,000 kWh. Since one kg of hydrogen has an energy content (lower heating value) of 33.31 kWh, then the SMR must produce 65,000/33.31 = 1951 kg/day of hydrogen to run the stationary fuel cell system.

---

[39]The Fuel Cell Energy DFC-300 molten carbonate fuel cell (MCFC) nominally has a 300-kW capacity, but is derated to 250 kW with the production of 125 kg/day of hydrogen.

[40]B. James, J. Moton, & W. Colella, "Mass production cost estimates of direct hydrogen PEM fuel cell system for transportation applications: 2014 update," by Strategic Analysis, Inc. of Arlington, Virginia.

Hydrogen Required for FCEVs

We assume in the model that vehicles travel an average of 11,876 miles per year, the US average for 2012. The National Laboratory-certified fuel economy[41] of the Toyota Highlander SUV FCEV was 68.3 miles/kg of hydrogen, so each FCEV requires approximately

11,876/68.3 = 173.9 kg of hydrogen per year or an average of approximately 0.476 kg/day.

The total hydrogen required then depends on how many FCEVs will be supported by each WWTP.

We assume that each WWTP station will support 60 FCEVs initially.

Note that a station that *supports* 60 FCEVs is *not* equivalent to a station that *refuels* 60 cars per day. For example, if we assume that a FCEV travels 11,876 miles per year (The US average for 2012) with a fuel economy of 68.3 miles/kg, then this FCEV would consume 0.48 kg/day as described above. If we further assume that the driver refuels his or her vehicle when it has used 85 % of the hydrogen (15 % remaining), then the driver will refill the car tanks approximately 35 times per year[42] or an average of 10.4 days between refills. In this case, a station supporting 60 FCEVs will refuel an average of only 5–6 cars per day, which is very low compared to some gasoline stations that refuel hundreds of cars per day.

This station that supports 60 FCEVs will on the average dispense 28.6 kg/day.

Total Hydrogen Required

The SMR system must therefore be designed to produce over 1980 kg/day (1951 kg/day for the stationary 1.3 MW fuel cell system plus 28.6 kg/day for FCEV refueling.)

WWTP System Cost Estimates

The WWTP hydrogen station will have all the subsystems shown in Fig. E.1 plus heat exchangers to recover waste heat from the stationary fuel cell and SMR systems.

[41]K. Wipke, D. Anton & S. Sprik, "Evaluation of range estimates for Toyota FCV-adv under open road driving conditions," National Renewable Energy and Savannah River National Laboratories, SRNS-STI-2009-00446, August, 2009.

[42]Assuming a hydrogen tank designed to hold 5.9 kg for 400 miles range.

The Anaerobic Digester

We assume that the anaerobic digester is built as part of the WWTP

Anaerobic Digester Gas Cleanup System

Anaerobic digester gas (ADG) is typically composed of 50–60 % methane (CH_4), 38–48 % carbon dioxide (CO_2), with traces of inert gases such as nitrogen. But ADG also contains H_2S and siloxanes that must be removed to avoid damaging downstream equipment. After removing the H_2S and siloxanes, the percentage of methane should be increased using either a membrane separator or a PSA system that uses adsorbents that remove much of the CO_2 while letting the smaller methane molecules pass through. The gas flow must be reversed periodically once the adsorbents become saturated with CO_2. Multiple adsorbent beds are used, so that one or more beds can be producing high concentrations of methane, while other beds are being cleaned of CO_2 by reverse flow. The Eastern Research Group[43] estimated that the cost of gas fuel treatment for a 300-kW fuel cell plant power by ADG would be approximately $147,000. We use this estimate in the model with a 0.6 power scaling factor for other sizes of FC plants:

$$\text{Cost}_{\text{gas clean-up}} = \$147,000\left[\text{Capacity}_{\text{FC}}(\text{kW})/300\right]^{0.6}$$

Heat Recovery System

The Eastern Research Group[17] also estimated a cost of $23,200 for a heat recovery system for the 300-kW fuel cell plant. So we assumed a heat recovery system cost of

$$\text{Cost}_{\text{heat recovery}} = \$23,200 \times \left[\text{Capacity}_{\text{FC}}(\text{kW})/300\right]^{0.6}$$

Steam Methane Reformer System

A SMR converts methane (the main constituent of natural gas) and water into primarily hydrogen and carbon dioxide. An SMR can also convert the methane in ADG into hydrogen and carbon dioxide. A commercial SMR such as the one my company, H2Gen, developed and sold also contains a gas filter to remove H_2S from the incoming gas stream and a compressor to raise the pressure of the methane into a range that is effective for reforming, nominally 13.8 bar. Finally, the SMR systems that we developed include a built-in PSA system to filter out all gases except hydrogen, producing an ultra-pure hydrogen stream suitable for powering PEM fuel cells. The estimated cost of such an SMR system is shown in

[43] Eastern Research Group, "Opportunities for and Benefits of combined heat and power at wastewater treatment facilities," prepared for the U.S. Environmental Protection Agency, December 2006.

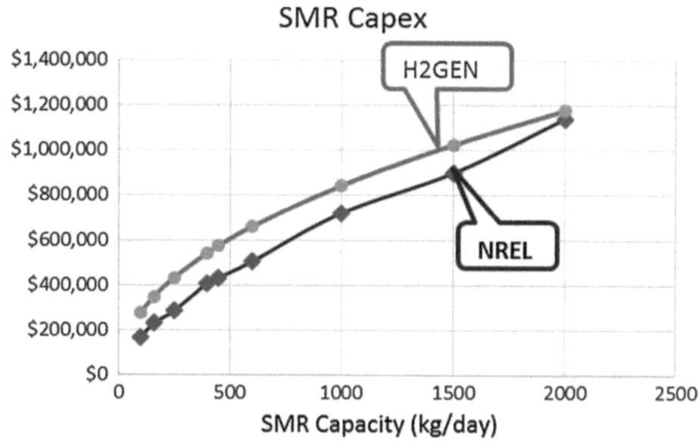

Fig. E.4 Two estimates of SMR capital cost as a function of plant capacity in kg/day

Table E.3 Markup factor to convert fuel cell cost to price

	Rate (%)	Markup
Profit	15.0	1.15
Non-recurring engineering	8.0	1.08
Warranty	20.0	1.2
Contingency	5.0	1.05
NET markup		1.56

Fig. E.4, along with an estimate made by the National Renewable Energy Laboratory (NREL).[44] A curve fit to the higher H2Gen estimated SMR system cost[45] which we use in the model is given by

$$\text{Cost}_{smr} = \$30360 \times C\,(\text{kg/day})^{0.48}, \text{where "C" is the capacity of the SMR in kg/day.}$$

PEM Stationary Fuel Cell System Cost

For the purposes of this calculation, we assume that the fuel cell companies building bus fuel cell systems are able to produce 160-kW systems for approximately $58,462 as projected by James et al., in production rates of 400 systems per year, or $365/kW at the cost level.[46] This cost estimate must be marked up to account for profit, warranty, non-recurring engineering charges, and a 5 % contingency as summarized in Table E.3. The total cost markup is by a factor of 1.56 or a fuel cell

[44]M. Melaina & M. Penev, "Hydrogen Station Cost Estimates: comparing hydrogen station cost calculation results with other recent estimates," NREL Report NREL/TP-5400-56412, September, 2013

[45]Both the NREL and H3Gen cost estimates do not include a hydrogen gas precooling system that is required to fill 700-bar FCEV car tanks; a refrigeration system is added separately below.

[46]B James, J. Moton, & W. Colella, "Mass production cost estimates of direct hydrogen PEM fuel cell system for transportation applications: 2014 update," by Strategic Analysis, Inc. of Arlington, Virginia.

Table E.4 Estimated salvage value for refurbished 160-kW fuel cell systems

		Salvage %	Salvage $	Salvage %
Stamped bipolar plates	$1139	50	$570	
MEAs:			$ -	
Membranes	$6988	0	$ -	
Catalyst ink	$4847	90	$4362	
GDLs	$5695	0	$ -	
M&E cutting and slitting	$13	0	$ -	
MEA Gaskets	$759	0	$ -	
Coolant gaskets	$175	0	$ -	
End gaskets (screen printing)	$1	0	$ -	
End plates	$133	100	$133	
Current collectors	$15	90	$14	
Compression bands	$16	0	$ -	
Stack insulation housing	$147	0	$ -	
Stack assembly	$139	0	$ -	
Stack conditioning	$389	0	$ -	
Stack total	$20,456		$5078	24.8
2 stacks	$40,912		$10,156.60	
Stack $/kW	$255.70			
Balance of plant				
Air loop	$7499	100	$7499	
Hunidifier and water recovery	$1219	50	$610	
High-temperature coolant loop	$1729	95	$1643	
Low-temperature coolant loop	$217	95	$206	
Fuel loop	$950	100	$950	
System controller	$533	95	$506	
Sensors	$4155	95	$3947	
Miscellaneous	$909	95	$864	
BOP total	$17,211		$16,224	94.3
Assembly and Testing	$339	0	$ -	
Total system	$58,462	$ -	$26,381	45.1

system *price* of $\$365 \times 1.56 = \$572$/kW. The 1.3 MW stationary PEM FC system that we are proposing in this model will therefore cost approximately $743,343.

We assume that these stationary fuel cell systems will have a life of 40,000 h,[47] which means that they must be replaced every 4–5 years. However, the cost of replacement will be less than the initial price since many of the parts can be salvaged each time a fuel cell system is refurbished. The net salvage value is estimated at approximately 45 % or $26,381 as shown in Table E.4. So the cost to

[47]M. Jimenez, "Smart electricity networks based on large integration of renewable resources and distributed generation," Kassel University Press, 2007.

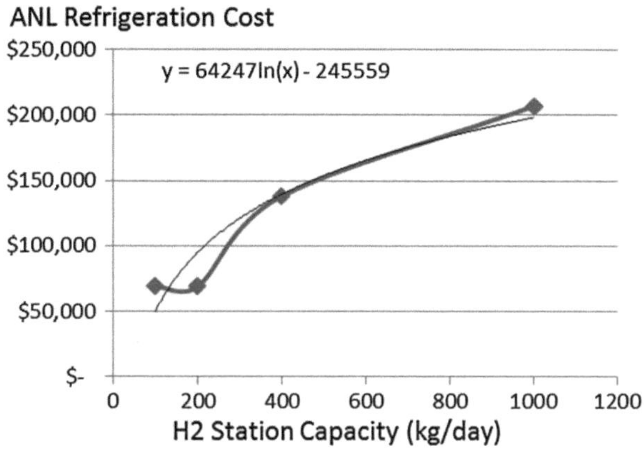

Fig. E.5 Cost of hydrogen refrigeration for filling 700-bar vehicle tanks, estimated by the Argonne National Laboratory

replace the fuel cells every 4 to 5 years is 55 % of the original price of $743,343 or approximately $407,909.

In addition, the model includes an annual operation and maintenance budget of 5.5 % of the initial price of the fuel cell system or approximately $40,900 per year; over the four years between fuel cell replacements, then, the project spends an additional $163,500 over the $407,900 spent to refurbish the fuel cell systems.

Hydrogen Cooling System

Hydrogen must be cooled to avoid overheating[48] while filling the FCEV car tanks at 700 bar, the pressure now used by most auto companies for their FCEVs. The cost of the refrigeration system as a function of station capacity is estimated by the Argonne National laboratory[49] as shown in Fig. E.5. We use the logarithmic curve fit to estimate the cost of the cooling system for hydrogen flows greater than 250 kg/day and $69,000 each for lower capacity stations.[50]

[48]This overheating is not a safety issue, but it does preclude the complete filling of a hydrogen tank since the warm hydrogen expands and takes up more space than the design quantity of hydrogen at room temperature.

[49]K. Reddi, A. Elgowainy, & E. Sutherland, "Hydrogen refueling station compression and storage optimization with tube-trailer deliveries," International Journal of Hydrogen Energy, 39 (2014), pp. 19164–19181.

[50]The logarithmic curve fit to the ANL refrigeration cost curve goes negative for station capacities less than approximately 40 kg/day, while ANL estimated a cost of $69,000 for both 100 and 200 kg/day stations.

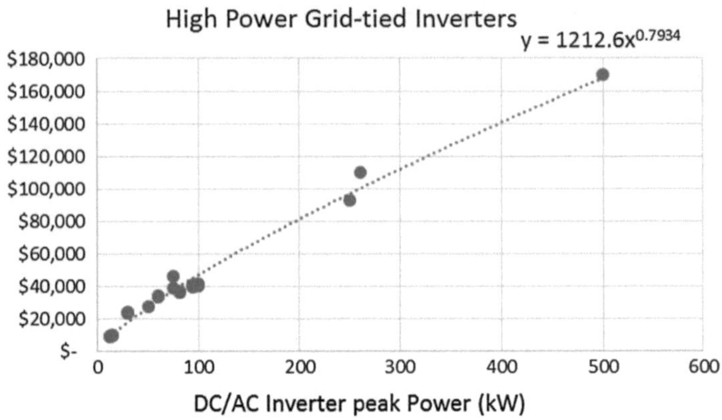

Fig. E.6 Cost of commercial DC to AC inverters as a function of power level and a power law fit to those data

Inverter

The direct current from the stationary fuel cell must be converted to alternating current with a DC to AC inverter. Figure E.6 shows a curve fit to DC to AC inverter costs taken from commercial inverter prices on the Internet in October 2014.

Hydrogen Compressors

Hydrogen compressors are needed to fill medium-pressure (110-bar) hydrogen storage tanks and to boost the pressure of hydrogen to the 850 bar to 950 bar range to fill 700-bar FCEV tanks.[51]

Hydrogen must be stored on-site at medium pressure to provide hydrogen for refueling vehicles that will arrive at irregular intervals. Elgowainy et al. at the Argonne National Laboratory, based on detailed analyses to minimize costs, recommend that low-pressure storage be provided for at least 15 % of the daily hydrogen load[52] or 4.3 kg of low pressure based on the daily consumption of 28.6 kg/day derived in Sect. 6.2 above.

[51]These two functions could be achieved with different stages of a single compressor, but we assume two separate compressors in this model which should be conservative relative to cost.

[52]A. Elgowainy (ANL), M. Mintz (ANL), B. Kelly(Nexant), M. Hooks(Tiax), & M. Paster(DOE), "Optimization of compression and storage requirements at hydrogen refueling stations," presented at the ASME pressure vessel and piping conference, Chicago, Illinois, July 27-31, 2008.

Table E.5 Compressor cost parameters from Yang and Ogden

Source	Base cost	Scale factor
	Cb	α
NAS-forecourt-current	$7237	0.8
NAS-forecourt-future	$2279	0.8
Yang and Ogden	$1888	0.9

Yang and Ogden at the Institute of Transportation Studies at the University of California at Davis[53] have analyzed various estimates of hydrogen compressor costs based on the compressor power required, given by:

$$P_{\text{comp}} = \text{QZRTN}\gamma/\left[\eta \times (\gamma - 1)\right] \times \{(P_{\text{o}}/P_{\text{i}})\hat{}\left[\gamma - 1\right]/\gamma\text{N}] - 1\} \quad \text{(E.4)}$$

where

- Q hydrogen flow rate (Nm^3/s)
- Z compressibility of hydrogen
- R universal gas constant =8.3146 J/Kmol
- T temperature (°K),
- N number of compressor stages,
- η adiabatic efficiency of compressor (65 % here)
- γ ratio of specific heats (1.41 for hydrogen)
- P_{o} output pressure (Pa), and
- P_{i} input (suction) pressure (Pa)

The system is designed to support 60 FCEVs, each of which consumes an average of 0.48 kg/day of hydrogen or 6.52 kg/h = 0.0201 Nm^3/s. We assume that the bulk of the hydrogen for refueling is stored at 110 bar, with a 3-tank high-pressure cascade with the highest pressure at 850 bar to 950 bar. The pressure out of the PSA is 200 psig or 13.8 bar. From Eq. E.4, the power required from the low-pressure compressor (13.79–110 bar) is 66.2 kW and the high-pressure compressor (110–850 bar) requires 38.7 kW for a total power draw of 105 kW.

Yang and Ogden[16] have also summarized the cost of compressors dependent on their power rating as shown in Table E.5,

$$\text{where } C_{\text{p}} = C_{\text{b}}\text{P(kW)}^{\alpha}$$

NAS refers to the 2004 National Academy of Sciences report on hydrogen.[54] They considered 2020 as "future" in this 2004 study, which will be here by the time large-scale deployments of FCEVs are completed, so we averaged the current and

[53] C. Yang & J. Ogden, "Hydrogen delivery and refueling infrastructure" the Institute for Transportation Studies, the University of California at Davis, provided for the California Energy Commission Advanced Energy Pathways (AEP) Project, June 2007.

[54] "The hydrogen economy: opportunities, costs, barriers, and R&D needs," Committee on alternatives and strategies for future hydrogen production and uses, National Research Council, ISBN 0-309-53067-7 (2004) Available at: http://www.nap.edu/catalog/10922/the-hydrogen-economy-opportunities-costs-barriers-and-rd-needs.

Table E.6 calculated compressor parameters and average costs based on the Yang and Ogden equations and NAS estimates

	H2 flow	kg/h	2.38	2.38
	H2 flow	Nm3/s	0.0074	0.0074
	#FCEVs	60	High pressure	Low pressure
	Pressure range (Bar)		110–850	13.8–110
	Cb	α	21.7 kW	131.5 kW
Source	Base cost	Scale factor		
NAS-forecourt-current	$7237	0.8	$84,868	$358,571
NAS-forecourt-future	$2279	0.8	$26,726	$112,917
Yang and Ogden	$1888	0.9	$30,119	$152,369
Average cost			$47,237	$207,952
Total compressor cost				$255,190

future NAS projections along with the Yang and Ogden estimates to arrive at an average compressor cost as shown in Table E.6.

The total estimated cost for the two compressors is $255,190 as shown in Table E.6. This calculation assumes an intermediate bulk storage pressure of 500 bar. Compressor costs could be reduced further with lower bulk storage pressures, but then, the bulk storage costs would increase along with the required storage volume.

Hydrogen Dispenser Cost

Some analysts have suggested dispenser costs in the $50,000 range, but the NREL independent peer review team[55] estimated that the dispenser would cost $165K as an optimistic case and $189K for their base case, due to the much higher operating pressures for hydrogen compared to natural gas compressor and the need to dispense cold hydrogen. We use the optimistic estimate of $165K for this model.

Electricity Consumption

The SMR and hydrogen compressors both consume electricity, which will reduce the electricity from the stationary PEM fuel cell that could be sold or used to displace electricity previously purchased from the local utility.

[55] G. Parks, R. Boyd, J. Cornish and R. Remick, "Hydrogen Station Compression, Storage and Dispensing: Technical status and costs," the National Renewable Energy Laboratory independent peer review team, NREL/ BK-6A10-58564, May 2014.

SMR Electricity Consumption

The HGM-10,000 SMR system[56] consumes 1.48 kWh/kg of hydrogen produced. Most of this electricity is used in the natural gas compressor. Compressor power is directly proportional to the gas flow rate as shown by Eq. E.4. So this electricity per kg value will increase with increased hydrogen flow. We assume that the SMR compressor will consume:

$$\mathrm{SMR_{electricity\ consumption}} = 1.48\ \mathrm{kWh/kg}\ \mathrm{C(kg/day)}/567\ \mathrm{kg/day},$$

where C(kg/day) is the capacity of the SMR.[57] For the base case of a 1.3-MW stationary PEM fuel cell system, the required hydrogen flow is approximately 1951 kg/day plus 23.8 kg/day for the FCEVs, which increases the SMR electrical consumption to 5.17 kWh/kg.

The SMR then consumes approximately 10,276 kWh/day or 3.62 million kWh/year with a 96 % system annual availability factor.

Hydrogen Compressor Electricity Consumption

The hydrogen compressors only compress the hydrogen needed for refueling FCEVs. The hydrogen used to run the stationary PEM fuel cell system is fed directly to the fuel cell; there is no need to store hydrogen for the stationary fuel cell system.[58] So compressors are only needed for the hydrogen used for FCEV refueling. The model assumes that the hydrogen fueling station is designed to accommodate 60 FCEVs. Each FCEV consumes an average of 0.48 kg/day, so a station designed to support 60 FCEVs will require a hydrogen flow of 28.6 kg/day on the average. The low-pressure compressor must operate continuously to fill the 500-bar storage tanks at a power level of 131.5kW or an annual electricity demand of 184,400 kWh/year. The high-pressure compressor (500–900 bar) requires 21.7 kW, but only operates 105 h per year when one of the 60 FCEVs is being filled, or a total electricity demand of 2210 kWh /year for a total annual demand of 86,600 kWh/year from both compressors, which is just 1.7 % of the annual 1.3 MW FC output of 11.05 million kWh/year.

[56]The HGM-10,000 is the commercial SMR developed by H2Gen, Inc. that produced 10,000 standard cubic feet of hydrogen of hydrogen per hour or approximately 567 kg/day.

[57]To the degree that some of the electricity in the SMR is used for controls and devices that do not increase with hydrogen flow rate, this adjustment of electricity consumption is conservative.

[58]Although some WWTP operators might want to store some hydrogen to serve as an emergency backup system in case of a power outage. Hydrogen could also be stored to be used to cover annual maintenance downtime on the SMR system.

Table E.7 Summary of hydrogen station costs for a WWTP in thousands of dollars

Gas pretreatment	$354.3
1.3-MW PEM FC	$743.3
SMR system	$1171.2
Low-pressure compressor	$208.0
High-pressure compressor	$47.2
Low-pressure storage	$10.6
High-pressure Cascade storage	$44.5
Refrigeration system	$69.0
Dispenser	$165.0
DC/AC inverter	$358.3
Heat recovery system	$55.9
Electrical and controls	$161.3
Mechanical and piping	$166.7
Total Capex	$3555.39
Installation costs (20 %)	$711.1
Total installed costs	$4266.5

Installation Costs

The cost of installation is a significant fraction of the total project costs. The NREL technical peer review team[18] recommended installation costs between 20 % (Optimistic case) and 30 % (base case) of the project capital costs. We use an installation factor of 20 % in the model.

Summary Costs

The total system capital costs of $3.6 million are summarized in Table E.7, with a total installed cost of $4.27 million.

FSC
www.fsc.org
MIX
Papier aus verantwortungsvollen Quellen
Paper from responsible sources
FSC® C105338

If you have any concerns about our products,
you can contact us on
ProductSafety@springernature.com

In case Publisher is established outside the EU,
the EU authorized representative is:
Springer Nature Customer Service Center GmbH
Europaplatz 3, 69115 Heidelberg, Germany

Printed by Libri Plureos GmbH
in Hamburg, Germany